Wörterbuch

der

Schleif- und Poliertechnik

von

Senator h. c. B. Kleinschmidt

Teil I

Deutsch — Englisch

Teil II

Englisch — Deutsch

Mit 25 Abbildungen

Berlin W 35

TECHNISCHER VERLAG HERBERT CRAM

1952

Fachausdrücke

der

Schleif- und Poliertechnik

Teil I

Deutsch-Englisch

Vorwort

In jeder Industrie werden viele Ausdrücke und Redewendungen verwendet, die nicht in den gewöhnlichen technischen Wörterbüchern enthalten sind. Man findet sie nur in einem Fachwörterbuch. Die Arbeit des Uebersetzers wird auch dadurch sehr erschwert, daß die Bedeutung vieler Sonderausdrücke ganz verschieden ist, je nachdem welches Verfahren, Stoff, Werkzeug usw. dabei Anwendung findet. Infolgedessen beruht der praktische Wert eines technischen Fachwörterbuches in erster Linie auf der mehr oder weniger großen Zahl von Fachausdrücken, also auf seiner Vollständigkeit, und vor allem auf der Genauigkeit ihrer Definition.

Das vorliegende Fachwörterbuch soll dem Schleif- und Polier-Fachmann, dem Verbraucher der vielartigen Schleif- und Poliermittel, dienen. Es enthält in erster Linie die Fachausdrücke der Schleif- und Poliermittel-Industrie, die wegen ihrer Neuheit und ihrer Sonderheit in den allgemeinen technischen Wörterbüchern nicht enthalten sind.

Senator h. c. B. Kleinschmidt

Reihenfolge des Fachwörterbuches

Abdrehapparat dresser
Abdrehapparat (Scheibe) metcalf wheel dresser
Abdrehapparat mit Abdrehstab stick dresser
abdrehen to dress
Abdrehen dressing
Abdrehen mit dem Diamanten diamond dressing
abdrehen, auf Maß to cut to size
Abdrehrädchen cutter
Abdrehvorrichtung, Schleifscheiben- grinding wheel dresser
Abdrehwerkzeug dressing tool, wheel dresser
Abfall waste
abgeschrägte Fläche beveled face
abgestuft stepped
abgestuft, geometrisch geometrically stepped
abgraten to burr. to fettle, to trim off
Abgratmaschine burring machine
Abgratscheibe snagging wheel, fettling wheel
Ablehren des Schleifsteines truing the grinding wheel - trimming the grinding wheel
Abmessung dimension
abnahmefertig o. k.; O. K.
Abnahmeprüfung inspection
abnehmen (bei der Revision) to pass inspection
abnutzen, sich to wear out
Abnutzung wear
abreiben to abrade
abreiben, mit Bimsstein to pounce
abrichten to true, to trim, to dress
Abrichten truing, truing up, trimming dressing
Abricht-Praxis dressing practice
Abricht-Zeit dressing time
Abrichtvorrichtung truing device
Absatzschleifpapier heel scouring paper
absaugen to exhaust
Absaugung exhaustion
abschaben to abrade
abschleifen to abrasive, to rub down to surface
Abschleifen rubbing down
Abschleifen des Absatzes heel scouring
Abschleifung abrasion
Abschluß, wasserdichter water seal
abspannen to loosen, to unclamp
Abstechmaschine cutting-off machine
Abstechwerkzeug cutting-off tool
abstufen to grade
Abwälz-Verfahren self-generating method
Abweichung deviation, variation
abziehen to hone
Abziehstein oilstone, sharpening stone rubber, hone, whet stone
Abziehstein in Messerklingenform slip stone
Abziehstein für allgemeine Zwecke general purpose stone
Abziehstein für ärztliche Instrumente instrument hone
Abziehstein für Tuchschneidemesser cloth cutters' stone
Abziehwerkzeug wheel dresser
allgemeine Schleifzwecke general grinding
allgemeine Werkstatts-Zwecke general shop use
Aluminium aluminum
Aluminiumoxyd aluminum oxide
anfertigen, genau die gleiche Scheibe - to duplicate a wheel
Anglerstein hook stone, hook hone
anheften to stick

anlassen to start, to temper
Anpreßdruck gegen force acting against
anreißen to lay out, to mark off
Anschlag dog
Anschlag, verstellbarer stop dog, stop gauge, stop
anschleifen, leicht to touch up
Antrieb drive
Antriebsscheibe für das Werkstück live spindle driving pulley
Antriebsscheibe für tote Spitze dead center pulley
Antriebswelle driving shaft
anzeigen to gage, to gauge
Anzeiger index
anziehen to tighten
Arbeiter operator
Arbeitsgeschwindigkeit operating speed
Arbeitsmaschine, leichte light duty machine
Arbeitsmaschine, schwere heavy duty machine
Arbeitsstück work piece
Arbeitstakt pace
Arkansas-Stein arkansas stone
Arbeitsvorgang operation
ätzen to attack
ätzpolieren to attack-polish, to etch-polish
Aufkugeln pressfinish
Auflage, Schleif- grinding rest
aufputzen to furbish up
aufrecht endwise
Aufschrift label
aufspannen to clamp, to set
Aufspannvorrichtung chuck
Aufspannvorrichtung. magnetische magnetic chuck
Aufstellung (einer Maschine) foundation set up
Wörterbuch **Spalte 2**
auftragen to build up
auftragen, etwas auf eine Scheibe to charge a wheel with
ausbalanzieren to balance
ausbessern to fettle
auseinandernehmen to disassemble
ausfeuern lassen (Schleifscheibe) to spark out
ausfluchten to line up with
ausgebleit bushed
ausgerichtet zu to be in line with
ausgespart recessed
auskehlen to furrow
Auskehlen (Steinindustrie) furrowing
Ausladung overhang
auslegen to lay out
Auslösung, selbsttätige automatic release
ausprobieren —cut and try
Ausrichtung alignment
ausrücken to knock off, to throw out
Ausrüstung equipment, outfit
Ausschlag deviation
ausschleifen to grind internally, to grind out
Ausschleifen internal grinding, grinding out
außerachsig off center
äußerlich external
Aussparung recess
Aussparung, Scheibe mit cupped wheel
austauschbar interchangeable
Austauschbarkeit duplication
Auswahl der Bindung selection of bonding
auswechselbarer Steinhalter refillable stoneholder
auswerfen to knock out
auswuchten to balance
automatisch automatic (ally)
Auto-Stein auto-stone
Axtstein axe stone, lathers stone

ballig crowned
Balligschleifeinrichtung cambering attachment
Balligschleifeinrichtung zum Schleifen von Walzen cambering attachment for grinding slightly convex rolls
Band belt
Bandpoliermaschine belt polishing machine
Bandschleifmaschine belt grinder, belt grinding machine
Bankschleifmaschine bench grinding machine
Bandschleifmaschine für schwere Stücke heavy duty bench grinder
Bankstein bench stone
basicher Ton aluminous clay
Baumwollscheibe, ungenäht, zum Polieren cotton dolly, cotton mop
Bauxit Bauxite

Beanspruchung stress, strain
bearbeiten to dress
bearbeitet finished
bedienen to operate
Bedienungsmann operator
befestigen to fasten, to secure
befilzt felted
Beistellung in-feed
beizen to pickle
Belag coating
Belag, einfacher single coating
Belag, doppelter double coating
Belastung load
belegen to coat, to set up
belegt faced
belegt, doppelseitig double faced
belegt, einseitig single faced
bersten to crack
Bersten der Schleifscheibe rupture or bursting of the grinding wheel
Berührungsfläche zwischen Schleifscheibe und Werkstück area of grinding contact
bestoßen to burr, to trim off
bestreuen to coat
Bettschlitten grinding carriage
Bewegung motion, movement
Bewegung, hin und hergehende alternate, reciprocated motion
biegen to curl
Bildhauer-Abziehstein carvers' slip
Bindemittel bonding agent, bonding material
Bindung (der Schleifscheibe) bond
Bindung, keramische vitrified bond
bimsen -to rub with pumice, to pumi cate
Bimskissen pad
Bimsstein pumice, pumice brick, pumice stone
Bimsstein-Pulver pounce
blank gewordene Scheibe glazed wheel
Blankputzen scouring
Blatt (Bogen) sheet
Blattware sheet goods
Blech plate. sheet
Block block
Blutstein bloodstone
Bodenstärke (einer Topfscheibe) thickneß back
Bogen (Blatt) sheet
Bogen (Krümmung) arc
Bohrer-Schleifmaschine drill grinder, drill grinding machine
Bohrfutter chuck, drill, socket
Bohrung bearing, bore, boring, hole, hollow
Bohrung, konische taper hole
Bolzen bolt, pin, stud
brechen to crush
Brecher crusher
Brechwalzwerk crusher
Breite width
Bremstrommel brake drum
Bremstrommel-Schleifmaschine brake drum grinder
brennen to burn
Brennkapsel sager
Brennkapselton sagers' clay
Brennofen kiln, vitrifying kiln, oven
Brille center rest
Bruch breakage, break down, fracture
Bruch der Schleifscheibe rupture of the grinding wheel
Buchse bush, bushing, shell, sleeve
Buchsenschleifmaschine internal grinder
Bund collar, shoulder
bündig flush
bürsten to brush
Bürste brush

chemische Zusammensetzung chemical composition
Chromoxyd chrome oxide

Dauerhaftigkeit durability
Dehnbarkeit ductility
Diamant diamond
Diamanthalter diamond tool holder
Diamantschleifwerkzeug diamond lap
Diamantstaub diamond dust
Diamantwerkzeug diamond tool
Diatomeen-Erde diatomaceous earth
dichtes Gefüge dense structure
dichte Streuung close coating
Dicke gage, thickness
doppel . . . twin . . .
Doppelschleifscheiben, das Schleifen mit twin wheel grinding
Doppeltopfscheibe double cup wheel, doube tub wheel
Dorn arbor, arbour, mandrel, plug

Dorn, Polier- buffing cone, polishing cone
Dorn, Schleif- plug lap
Dorn,, Schleifscheiben- grinding wheel arbour
Draht wire
Drahtgewebe-Scheibe (mit Drahtgewebe-einlage) wire-web wheel
Drahtbürste wire brush
Drehbankspitzen-Schleifapparat lathe center grinding apparatus
drehbar pivoted, swivable
drehbar pivoted, swivelable
drehen, sich to revolve, to rotate, to swivel, to turn
Dreherei (Fertigdreherei der gebrannten Scheiben) lathe room
Drehfutter lathe chuck
Drehung rotation
Drehzapfen journal, pivot
Dreibackenfutter chuck
Dreikantfeile triangular stick
Druckauflage pressure rest
Druckluft-Spannfutter pneumatic chuck
Durchmesser diameter
durchlässig porous
Durchlässigkeit porosity

eben even, flat
ebenschleifen to grind even, even grinding
Ebene level
Edelsteinschliff gem grinding
egalisieren to true up
Eindrehung groove
Einschleifen grinding in
eichen to adjust
einfach plain
einfache glatte Scheibe plain straight wheel.
Einfache (Rund-) Schleifmaschine plain grinder
Eingekapselte Schleifmaschine self-contained grinder
eingreifen to engage, to mesh
Eingriff contact, engagement
einkerben to notch
Einkitten der Schleifscheibe cementing - in of the grinding wheel
einpassen to adjust
einrichten to set up
Einrichtung setting up, engagement
einrücken to engage, to throw in
Einsatz socket
Einscheiben-Antrieb single pulley drive
einschleifen (Ventile) to reface
Einschleifen (Glas) stoppering
Einschleifpaste grinding compound, lapping compound
Einschnitt groove
einsetzen to set, to insert
einspannen to chuck, to clamp
Einständer-Flächenschleifmaschine open side surface grinder
Einstech-Verfahren in-feed method
einstellbar adjustable
einstellen to adjust, to locate, to register to set
Einstellung adjustment, location
eisenhaltig ferreous
elastisch elastic
elektrisch electric
elektrische Schleifmaschine electric grinder
Elektrocorund aluminous abrasive, aluminum oxide
Elektrokorund-Polierkörner aluminous polishing grains
elektromagnetisch electro magnetic
endlos endless
endloses Band endless belt
Endmaß precision gauge block
Endmaß, sphärisches end measuring block or rod
Englische Erde rottenstone
Englisch Rot caput mortuum, colcothar, English red
entfernen to take out
entgegengesetzt zum Uhrzeigersinn counterclockwise
Entlüftung vent
Erschütterung chatter, vibration
Etikett washer, label, tag
Exhaustor exhauster
Exzenter cam, excentric disc
exzentrisch excentric

Fabrikat make of machine
Facette bevel
facettieren to bevel, to cut facets
Facetten feinschleifen to fine bevels

faltige Verbindung puckered joint
Fassonabziehstein für Bildhauer carvers' slip
Feder spring
federnd spring actuated
Feile file, stick
Feile, konische point
feilen to file
Feineinstellung, selbsttätige automatic fine feed
feine Schicht film
feiner Ueberzug film
Feinheit fineness
Feinhieb finishing cut
Feinkörnig fine grained
Feinkörnigkeit fineness
Feinmachen - Feinschleifen
Feinpolieren buffing
Feinschleifen (Glasindustrie) smoothing
Feinschleifpapier finishing paper, pouncing paper
Feinschleifprocess (Steinindustrie) ironing
Feinschliff finish, smoothing
Feinschnitt finishing cut
Feldspat feldspat, felspar
Fertigläppen finish lapping
fertigmachen to finish
fertigpolieren to finish polish
Fertigschliff finish grinding, finishing grinding
festsetzen, sich to stick
festspannen to lock
feststehend stationary
feststehende Maschine rigid machine
feststehende Spitze dead center, false center
feststellen to register
Fett fat, grease
Fett, konsistentes non fluid oil
feuchtigkeitsbeständig moisture proof
feuern to spark
Feuerstein firestone
Filz felt
Filzkissen felt pad
Filzpolierband felt polishing belt
Filzpolierscheibe felt polishing disc
Filzscheibe felt washer
Filzunterlegscheibe felt washer
Fingerformverbindung interlocking splice
flachfräsen to plane
Flachscheibe plain wheel
Flächenschleifen to surface, surface grinding
Flächenschleifmaschine surface grinding machine
surface grinder, disc sander
Flansch flange
Flatterscheibe floating wheel
fliegende Schleifscheibe overhung grinding wheel
Fliehkraft centrifugal force
Flint flint
Flintschleifpapier flint paper
Flügelmutter fly nut, thumb nut
Flußspat Derbyshire spar
Flüssigkeitsgetriebe control oil gear pump
Form geben, einer Scheibe tho shape a wheel
Formscheibe formed wheel
formschleifen (Marmorindustrie) to mould
Formschleifen moulding, form grinding
Fräserschleifmaschine cutter grinder
Freihandschleifen free hand grinding
freischneiden to cut cleanly
Freischneiden free cutting
Freischnittwinkel clearance angle
Fundament foundation
funken to spark
Funkengarbe ray of sparks
Funkenprobe spark test
Fußboden-Flächenschleifmaschine floer surfacing grinder
Fußbodenschleifmaschine floor grinding machine
Fühllehre feeler
Führung bearing, guide, way
Führungsbahn guide way
Führungsbuxe guide box
Füller (Porenfüller beim Polieren) filler

Galvanisieren electroplating
Gang, in - setzen to start
Gang, ruhiger - steady running
Gang, toter - backlash, end play, lost motion
Gangtiefe depht of thread
Gefüge structure, texture
Gefüge, dichtes dense structure
Gegendruck end thrust
Gegengewicht counterweight

gegossene Scheibe puddled wheel
gehärtet hardened
Gehäuse housing
gekörnt grained
gekröpft off-set
gekrümmt curved
genähte Baumwoll-Schwabbelscheibe stiched cotton buff
genau, vollkommen dead true
genau rund laufen to run true
Genauigkeit trueness
Genauigkeitsgrenze limit of accuracy, degree of accuracy
gepreßte Scheibe pressed wheel
gerade even, straight
Geradschliff der Zähne sharpening the theet square across
gerade Topfscheibe straight cup wheel
Geradheit straightness
gerillt fluted
Geschwindigkeit speed, rate, velocity
Geschwindigkeit einhalten to check the speed
Geschwindigkeitsbereich speed range
Gestell frame
Getriebe gear, mechanism, movement
Getriebekasten speed box
Getrieberäderkasten speed change box
Gewebestoff für Polierscheiben fabric for wheels
Gewinde thread
Gewindebohrer screw tap
Gewindelehre thread gauge
Gewindeschliff thread grindling
Gewindeschneidwerkzeug screw cutting tool
Glanz glosse, luster, lustre
glanzschleifen to buff, to burnish, to glaze, to polish
glashart brittle
Glas glass
Glaspapier glass paper
Glasstöpsel einschleifen to grind in a stopper
Glasstöpsel-Einschleifen stoppering work
glatt smooth, plain
Glattmachen (Glasindustrie) facing off
glattgeschliffene Oberfläche smooth ground surface
glattschleifen (Betonböden) to surface
Glattschleifen (Glas) smoothing
Glattwerden glazing
Glänzen burnishing
glätten to brighten, to burnish
to planish, to smooth
Glätten burnishing
Glätten (Betonböden) surfacing, ironing
Glätten (3. Arbeitsgang beim Glasschleifen) smoothing
Glätter furbisher
Glättscheibe (Sonderscheibe für die Steinindustrie) ironing wheel
Glättstahl burnisher
Glättzahn polishing iron
gleichmäßig machen (Glasindustrie) facing off
Gleichmäßigkeit uniformity
Gleichmäßigkeit der Körnung uniformity of grit
Gleichstrom D. C. (direct current)
Gleitbahn slide
gleiten to slide
glühen to anneal
Grad grade
Gradschliff der Zähne (einer Säge) sharpening the teeth square across
Gradteilung graduation
Granat garnet
Granatkombination garnet combination
Granatpapier garnet paper
Granatschleifleinen garnet cloth
Grat barb, burr, seam
Graveurstichel engravers' pencil point
Grenzlehre limit gauge
Grenzmaß, oberes high limit
Grenzmaß, unteres low limit
Grenzrachenlehre caliper gauge
Grobheit coarseness
grobkörnig coarse grained
grobschleifen to grind rough, to grind coarse
Grobschliff roughing, first grinding
Grobschliffscheibe fettling wheel
Gummi rubber
Gummi-Bindung rubber bond
Gummischeibe rubber washer, rubber wheel
Gummi-Schleifscheibe rubber grinding wheel
Guß putzen snagging
Gußputzmaschine snagging machine
Gußputzscheibe snagging wheel
Guttapercha rubber
Güte grade, fineness, quality
Güte des Fertigschliffs quality of finish

Haarriß (im Stahl) check
Hacken-Schleifmaschine hoe grinder
halbselbsttätig semi-automatic
Halter für den Wasserfänger water guard support
Haltevorrichtung locating device
Handauflage hand rest
Handgriff handle
Handrad hand wheel
Handschleifen hand grinding
Handschleifmaschine hand grinder, portable grinder
Handschleifstein rubber, slip
Handsupport hand rest
Harnstoffharz urea formaldehyd resin
Hartzerkleinerungsmaschine crusher
Harz resin
Harzbindung resin bond
hängende Schleifmaschine over head grinding machine
Härte grade, hardness
Härtebericht grading report
Härtebestimmung grading
Härtegrad degree of hardness, temper
härten to temper
Härteprüfmaschine hardness testing machine
Härteprüfung hardness testing
Härteriß crack
Härteskala grade scale, scale of hardness
Hauptwelle line shaft
hellputzen high polish
Herstellungszeit set up time
Hinterschleifen to back off, to relief grind
Hinterschleifeinrichtung relief grinding attachment
Hinterschliff clearance
hin- und herbewegen to reciprocate
hin- und hergehende Bewegung alternate motion
Hobeleisen-Klappe plane-iron-cap
Hobelmesser-Schleifmaschine grinding machine for plane irons
hochglanzpolieren to buff, to burnish, to high polish
Hochglanzpolitur brilliant polish, high polish
hochtourige Scheibe high speed wheel
hohlgeschliffen clearance ground, hollow ground
Hohlkehle flute, fluting, hollow
Hohlmeißelstein gouge stone
Hohlmeißelstein mit abgerundeter Kante round edge gouge stone
hohlschleifen to grind clearance, to grind hollow
Hohlschleifen clearance grinding, internal grinding
Hohlschleifmaschine internal grinding machine
Holzbuchse wood bushing
Holzpolierscheibe box board lap
honen to hone
Honen honing
Honeinrichtung honing equipment
Honstein honing stick, honing stone
Holzschliffscheibe pulp grinding wheel
hydraulisch hydraulic

Innenschleifapparat internal grinding attachment
innenschleifen to grind internally
Innenschleifen internal grinding
Innenschleifmaschine internal grinder internal grinding machine
Innenschleifvorrichtung internal grinding fixture
Instandhaltung upkeep, up-keep
Ist-Maß actual-size

Kaliber caliber, groove
Kalibermaßstab caliper rule
Kaliberschleifmaschine sizing grinder
Kalk, Wiener Vienna lime
kanten to tilt
Kantenschleifen edging
Kanvas (Segeltuch) canvas
Kaolin china clay
Kapselton sagers' clay
Kardenzylinder-Schleifmaschine card-cylinder grinder
Kasten housing
Kegelneigung amount of taper
Kegelrad angular wheel, bevel gear
Kegelrad-Schleifmaschine bevel gear grinding machine
Kegelrollenlager taper roller bearing
Kegelschleifen cone grinding
Kehlmesser-Schleifmaschine moulding knife grinding machine

keramisch gebundene Scheibe vitrified wheel
Keramische Bindung vitrified bond
Kerbe nick, score
kerben to score
Kieselgur diatomaceous earth
Kieselsäure silicia
Klang, reiner (bei der Scheibenprobe) clear ring
klarschleifen clear polish
klemmen to bind
Klemmhebel clamp
Klinge blade
Klumpen lump form
Knagge cam
Kochsalz common salt, sodium chloride
Koks coke
Kolbenringschleifmaschine piston ring grinding machine
Kolbenschieber piston valve
Kombinations-Abziehstein combination hone
Kombinations-Bankstein combination bench stone
Kombinations-Schleifpapier combination abrasive paper
konisch tapered
konische Bohrung taper hole
konische Feile point
konisch Scheibe tapered wheel
konische Tassenscheibe cone shaped cup wheel
Konischschleifen cone grinding
Konischschleifvorrichtung bevel grinding attachment
Konizität amount of taper
Kontrolle inspection
Kontrollehre reference gauge
Konus cone, taper
Konusschleifwalze cone grinding roll
konzentrisch concentric
Kopf, Kopfstück head
Korngröße grit size
Kornmarken grain marks
Kornsieberei grain grading department
korrosionsbeständige Legierungen corrosion resistant alloys
Korund corundum, crystalline alumina
Korund-Schleifmaterial aluminous abrasive
Kosten der Scheibenunterhaltung cost of wheel upkeep
körnen to granulate
Körner (Spitze) center point
Körnerpunkt center mark
Körnerspitzen-Schleifapparat center grinding attachment
körnig granular
Körnung grain, granulation, grit
Kraftbedarf HP required (horse power)
kratzen to scratch
Kratzen (einer stumpfen Schneide) pulling
Kratzenbelag-Schleifmaschine card clothing grinder
Kratzer scratch
Kreissäge circular saw
Kreuzsupport slide rest
kristallinisch crystalline
Kronenschleifstein face grinding wheel
kugelige Scheibe spherical wheel
Kugellager ball bearing
Kugellager-Käfig ball cage
Kugelpolieren tumbling polish
Kugelscheibe ball wheel
Kugelschleifen spherical grinding
Kugelschleifmaschine ball grinding machine
Kulisse link
Kulissensteuerung link motion
Kunstharz (Scheibenbindemittel) phenol formaldehyd resin
Kunstkorund artificial corundum
Kupferpolierscheibe copper burnishing wheel
Kurbelscheibe crank disc
Kurbelwelle crank shaft
Kurbelwellenschleifen crank shaft grinding
Kurbelwellenschleifeinrichtung spline shaft grinding att.
crank shaft grinding att.
Kurbelwellenschleifmaschine crank shaft grinder
Kurve curve
Kurvenscheibe cam
Kurvenscheibenteil, der keine Bewegung veranlaßt dwell of a cam
Küchenstein kitchen stone
Kühlmittel coolant, cooling medium, lubricant
kühl schleifen to grind cooly
kühl schneiden to cut cooly
Kühlwasser cooling water
künstliches Schleifmittel artificial abrasive, manufactured abrasive

Lack lacquer
lackieren to lacquer, to varnish
Lackpoliermaschine lacquer polishing machine
Lager bearing, bracket
Lager, glattes plain bearing
Lagerbuchse bushhing
Lagerung der Scheibe storage of the wheel
Lamelle multiple disc
Langsamgang slow motion
Länge, der - nach lenghtwise
länglich oblong
Längsbewegung longitudinal feed
Längsgang length feed
Längsvorschub der Schleifscheibe longitudinal traverse of the grinding wheel, travel of the grinding wheel
läppen to lap
Läppen lapping
Läppmaschine lapping machine
Lauf, ruhiger steady run
Laufen to run
laufen, genau rund to run true
Laufprobe running test
Lederspaltmesser-Schärfer skiving knife sharpener
Leergangsverlust idle loss
Leerlauf idle or lost time
legierter Stahl alloy steel
Legierung alloy
Lehre gage, gauge
Lehre, Schleif- grinding gauge
Lehre für das Schleifen von Spiralbohrern twist drill grinding gauge
lehren to gauge
Lehrenschleifmaschine caliber gauge grinder
Lehrenschliff gauge grinding
Lehrdorn plug gauge
Lehrring ring gauge
Lehrwerkzeug gage, gauge
leicht anschleifen to touch up
leichte Arbeitsmaschine light duty machine
Leim glue
leimen to glue, to cement
Leimpresse für Leinenscheiben disc grinder press
Leinen cloth
Leinenkissen cloth pad
Leinenrollen verschiedener Breite cloth in odd width rolls
Leinenscheibe cloth disc
Leinenscheibe, konische sander cone
Leinenscheibe, schwere heavy duty disc
Leinensparrolle cloth economy roll
Leistung capacity
Leistung (Produktion einer Maschine) output
Leistung (bestimmte Arbeit) job
Leistung, mittlere average output
Leistungsfähigkeit efficiency
Leistungsvermögen capacity
linksgängig left handed
Linsenränder schleifen to edge lenses
Loch hole
Lochlehre internal caliper gauge
lose Körner loose grains
loslassen to release
losspannen to loosen
Löschpapier blotting paper
Lösung solution
Lötflächenschleifmaschine lap grinder
Lünette center rest

Magnesit magnesite
magnetisches Spannfutter magnetic chuck
magnetische Spannvorrichtung magnetic holding device
Mantellinie surface outline
Marke label, brand
Maß dimension, measure, size
Maß, Abdrehen auf to cut to size
Maßschleifen grinding to gauge
precision grinding
Maßstab rate, rule
Maschenweite mesh
Maschine leichter Bauart light duty machine
Maschinenaufstellung machine foundation
Maschinenbett base, bed, bench
Maschinenfundament machine foundation
maschinengeschliffen machine ground
Maschinenläppen machine lapping
Maschinenmesser-Feile raper file
Maschinenmesserstein knife stone
Maschinenöl machine oil
Maschinenschliff machine grinding
Maschine schwerer Bauart heavy duty machine
Massenschliff high production grinding

massiv solid
Materialabnahme stock removal
Material entfernen to remove stock
Materialmenge amount of stock
das Material verbrennen to burn the stock
Materialvorschub stock feed
mattieren to semi finish, dull grinding
mattschleifen to frost, to make opaque by grinding
mattgeschliffen (Glas) dead ground, satin finish
Mechanismus device, mechanism
mehrschneidig multiblade
Mehrzylinder-Maschinenschliff multiple drum sanding
Mengenschliff high production grinding
Messeinrichtung sizing device
messen to gage, to gauge, to measure
Messer blade, knife
Messer, eingesetztes inserted blade
Messerkopf-Schleifapparat face milling cutter grinding attachment
Messerschärfer knife sharpener
Messerschleifen knife grinding
Messerschleifmaschine knife grinder
Messerschmiedewaren-Abziehstein cutlers' stone
Messing brass
Messuhr dial gauge, dial inidicator
Messung, eine — machen to take a measurement
Meßwerkzeug measuring tool
Metall hoher Festigkeit high tensile metal
Metall niedriger Festigkeit low tensile metal
Metallkreissäge circular saw for metal
Mikrometer micrometer
Mikrometerlehre micrometer caliper
Minuslehre go gauge
Mischmaschine mixing machine, mixer
Mischraum mixing room
mittelfeinkörnig medium grained
mittlere Leistung average output
Mitnehmer driver, dog
Mitnehmerspindel-Feststellstift live spindle locking pin
Modell model, pattern
Modul module
Montage foundation, assembly, set up
montieren to assemble, to erect
Montieren der Scheibe mounting of wheel
montierte Spitze mounted point
Motorantrieb motor drive
Motorantrieb, unmittelbarer direct motor drive
Muffe sleeve
Muster model, pattern
Mutter nut
Mutter, zweiteilige split nut

Nachbestellung repeat order
nachgiebig non-positive re-finish
nachpolieren second polishing
nachschleifen to regrind, to stone
nadelkristallförmig acicular
Nadelscheibe pin wheel
Naht seam
Napf cup, shell
naßpolieren to polish wet
naßschleifen to grind wet
Naßschleifmaschine wet grinder
Naßschleifscheibe, wet wheel
Naßschliff cool grinding, wet grinding
Naßschliff mit wasserfestem Schleifpapier wet rubbing operation
Naturstein natural stone
Nebenprodukt by product
Nennmaß basic size
neu (für unfertig) green
neue Scheibe green wheel
neu belegen, eine Scheibe to recoat a wheel
neuen Glanz geben to furbish up
neuschleifen to regrind
nichtrostender Stahl stainless steel
Nocken cam
Nockenschleifmaschine cam grinding machine
Nockenwellenschleifeinrichtung cam shaft grinding attachment
Nonius vernier
Norm standard
Normal . . . standard
Normallehre standard gauge
Normalform standard shape
Nut flute, groove, slot
Nuten grooving
Nutwelle splined shaft

Oberfläche abschleifen to surface
Oberflächenfestigkeit surface tenacity

Ofenplatten-Schleifmaschine stove-plate grinder
offene Streuung open coating
Oel oil
Oelbehälter oil reservoir
Oelfangschale oil pan
Oelhaut oil film
Oelstein oil stone
Oelzuführung oil supply
Oelzuleitung oil supply
ordnen to grade
Ordnung, in o. k. O. K.
Oszillierende Kurvenschleifmaschine oscillating curve grinding machine

Papier paper
Pappflansch blotter
Pappscheibe washer of blottingpaper
Parallelstück parallel
Pariser Rot Paris red
Passung fit
Paßstift dowel pin
Paßstück adapter
Paste paste
pendeln to oscillate
Perlenschleifmaschine pearl grinding machine
Petroleum kerosene, kerosine
Pfanne cup
Pflugschar-Schleifmaschine plow grinder
Phenolharz phenol formaldehyd resin
Planbewegung crossfeed
plandrehen to plane
Plangang cross feed
Planscheibe face plate
Planschleifen face grinding, plane grinding, surface grinding
Planschleifmaschine surface grinding machine, planertype surface grinder
Planschliff flat surface grinding
Platte plate
plattieren to plate
pliesten grinding with a set up wheel
Pließtscheibe set up wheel
Polierabteilung polishing department
Polierarbeit polishing work
Polierbank polishing lathe
Polier-Blech scotch trowel
Polierbüchse (mit Halter) polishing clamp
Polierdorn, polishing cone, buffing cone
Polier-Drücken presspolishing
Poliereisen burnisher, polishing iron, polishing tool
polieren, to brighten, buff, burnish furbish, glaze, gloss, lap, polish, stone
Polieren burnishing
Polierer burbisher, polisher
Poliererei polishing shop
Polierbürste polishing brush
Polierfeile polishing file
Polierfilz polishing felt
Polierfläche polishing surface
Polierhammer polishing hammer
Polierholz polishing stick
Polierholz, befilztes felted polisher
Polierkissen polishing pad
Polierleder buff-leather, strap polisher
Polierlöffel button sleeker.
Poliermaschine polishing machine
Poliermaschine, Schraubenkopf- screw polishing machine
Poliermaterial polishing material
Poliermittel polishing agent
Polierplatte polishing plate, polisher, glazer
Polierpulver polishing powder, buffing powder
Polierrad polishing wheel, buffing wheel
Polierriemen polishing belt, belt polisher
Polierrollen von Bohrungen bearing, sizing
Polierrollen von Zahnrädern burnishing
Polierrot crocus, rouge
Polierrot-Kuchen rouge cake
Polierstab polisher
Polierstahl burnisher, burnisher steel burnishing stick
Polierstein (Blutstein) polishing stone, burnishing stone
Polierschaber scraper-burnisher
Polierscheibe polishing wheel
Polierscheibe aus Baumwolle cotton dolly, cotton mop
Polierscheiben-Abdrehwerkzeug buff-stick
polierschleifen to rub
Poliertisch polishing table
Poliertrommel polishing barrel, pol. drum

Polierwerkzeug polishing tool, burnishing tool
Polierwachs polishing wax
Politur polishing, polish, gloss
Porenfüller (beim Polieren) filler
Porosität porosity
porös porous
Prägepolieren imprint polishing
Präzisionsschleifen accurate grinding, precision grinding
Präzisionsscheibe precision wheel
Preßluftschleifmaschine pneumatic grinding machine
Preßpolieren pressfinish
Preßverfahren press mould method
Probelauf safety speed test, trial run
Profil outline
Profilier-Vorrichtung profiling attachment
profilierte Scheibe profiled wheel
Prüfapparat testing apparatus
Prüfmaschine testing machine
Prüfraum inspection department
prüfen to inspect, to test
Prüfung test
Pulver powder, swarf
putzen to clean, to furbish
Putzer cleaner
Putztrommel tumbler, tumbler barrel, tumbling barrel, rattler, cleaning drum

Qualität grade, quality
Quarz quartz
quarzig quartzy
Quarzsand, quartzsand
quer crosswise
Querbewegung (des Schleifrades) crossfeed
Quergang (des Arbeitstisches) transverse motion
Querschlitten transverse slide, cross slide
Querschnitt cross section
Quervorschub cross-feed

Rachenlehre snap gauge
Radsatz-Schleifmsachine car-wheel grinder
Rahmen frame
Rand face
Randform shape of face
Randform der Scheibe wheel face
Randstärke (einer Topfscheibe) thickness rim
Randstärke (einer abgesetzten Scheibe) width of face
Rasiermesserstein razor hone
Rattern chatter
Rattermarke chatter mark
Rauhschleifen roughing, first grinding
rechter Winkel acute angle
rechtsgängig right handed
rechtwinklig bearbeiten to square up
Reduziereinsatz reducing socket, taper sleeve
Regulierungsscheibe (spitzenloses Schleifen) regulating wheel, feed wheel
regulieren to control, to regulate
Reibahle reamer
reiben to rub
reibend abradant
Reibscheibe friction disc
Reibschleifen honing, stoning
Reibung friction
Reihe set
Reihenfertigung duplicating work, repetition work
reiner Klang (bei der Scheibenprüfung) clear ring
Reinheit purity
reinigen to clean, to cleanse
Reitstock footstock
Reitstock-Feststellhebel clamping clever
Reitstockspitze center point, footstock center, tailstock center
Reparaturwerkstatt repair shop, jobbing shop
Revolverkopf turret
richten to adjust
riefeln to flute
Riemen belt, strop
Riemen-Antrieb belt drive
Riemen getrieben belt driven
Riemenscheibe pulley
Riemenschleifmaschine belt grinder, belt grinding machine
Riemenspanner belt tightener
Riemenzug belt pull
Rille flute, groove
Rillenschleifen grinding out grooves
Ringscheibe ring wheel
Rinne groove
Riß (im Stahl) check, crack

Risse, feine (in Tiegeln) alligator cracks
Risse bekommen to crack
Rohlingsschlichtmaschine shaving machine
Rohr pipe, tube
Rohrleitung tubing
Rohschleifen to rough, coarse grinding
Rolle coil, rolls
rollen to curl
Rollenlager roller bearing
Rollenlager-Ring roller bearing cup
Rollieren rolling
Rost rust
rostfrei rustles, stainless
rostfreier Stahl stainless steel
Rücklauf reverse motion, return stroke
ruhiger Gang steady running
Rundfeile round stick
Rundheit roundness
Rundlaufen concentricity
Rundschleifapparat cylindrical grinding attachment
Rundschleifen circular grinding
Rundschleifmaschine circular grinding machine, cylindrical grinding machine, cylindrical grinder
Rundschliff circular grinding, cylindrical grinding
runzlige Verbindung puckered joint
Rutscherstein rubber, rubbing block, rub, rubbing brick, rubbing stone

Sand sand
Sandpapier sand paper
Sandpapierkonus sander cone
Sandpapiermaschine sanding machine, sand papering machine
Sandpapierschleifmaschine disc sander, sand papering machine
Sandstein sandstone
Sandstrahl-Gebläse sand blast apparatus
Satz set
sauberer Schnitt sheer cut
Sauberschleifen lapping
Sägemehl saw dust
Sägenschärfen saw gumming
Sägenschärfmaschine saw cutting machine, saw sharpening machine
Sägenschärfscheibe saw gumming wheel
Sägenschleifmaschine saw grinding machine
Sägenschleifscheibe saw gummer
Sägespäne scobs
säubern to clean up
Säule standard
Segerkegel pyrometric cone
Segment segment
Segmentscheibe segmental wheel, sectional wheel
Segmentscheiben-Spannfutter segmental wheel chuck
Segment-Schleifmaschine segmental grinding maschine
Selbstgang self feed
Selbstgang-Planzug power crossfeed
Selbstgang-Vorschub automatic feed
Selbsttätige Feineinstellung automatic fine feed
Setzstock back rest, steady rest
Sichelblatt-Schleifmaschine sickle-blade grinder
Sieb screen, sieve, sift
sieben to screen
Siebmaschine sifter, sifting machine, riddling machine
Siebnummerierung screen numbering
Siebweite mesh
Silberarbeiter-Abziehstein silversmith' stone
Siliziumkarbid carbide of silicon, silicon-carbide
sintern to sinter
Sitz fit
Räderschleifmaschine wheel grinder
sitzen to fit
Sorte grade
Sorteneinteilung grading
Span chip
Span abnehmen to remove stock
Spanbeseitigung chip clearance
Spanbildung chipping, chip formation
Spanwinkel rake angle
Spannbacke gripping jaw
spannen to tighten
Spannfutter chuck
Spannfutter magnetisches magnetic chuck
Spannfutter-Arbeit chuck work
Spannung stress, tension
Spannvorrichtung chuck, holding device
Sparrolle economy roll
Spat (Mineral) spar
Sperrklinke pawl
Sperrklinke für den Quervorschub crossfeed pawl

sphärisches Endmaß end measuring rod or block
spiegelglatt dead smooth
Spiegelpolitur ultra-finish
Spiel play, looseness
Spielraum clearance
Spindel arbor, arbour, shaft, spindle
Spindelgeschwindigkeit spindle speed
Spindelkopf spindle nose
Spindelkopf-Winkelstellungs-Anzeiger head stock index finger
Spindelloch arbour hole
Spindelstock head stock
Spindelstock-Drehplatte head stock base
Spindelstock-Spitze head stock center
Spiralbohrer twist drill
Spiralbohrer-Hinterschleifmaschine twist drill relief grinding machine
Spiralbohrer-Schleiflehre twist drill grinding gauge
Spiralbohrer-Schleifmaschine twist drill grinding machine
Spiralbohrer-Spitzenschleifapparat twist drill center grinding attachment
spiralig spiral
Spiralkurve spiral
Spitze, feststehende dead center
Spitze, umlaufende live center
Spitzenarbeit work between centers
Spitzendruck-Adjustierschraube tension adjusting knob
Spitzenentfernung distance beween centers
Spitzenhöhe height of centers
Spitzen, montierte mounted points
Spitzenschleifapparat center grinder
Spitzenschleifmaschine center grinding machine
Spitzenlose Rundschleifmaschine centerless grinding machine
spitzenloses Schleifen centerless grinding
Spitzgewinde angular thread
springen to crack
spröde brittle
Sprung crack
Stab rod
Stabilität rigidity
Stahlkernscheibe steel center wheel
Stahl, legierter alloy steel
Stahl, nichtrostender stainless steel
Stahl, rotstfreier rustless steel
Stahlsand steel shot
Stahlwolle steel wool
Stange rod
Ständer frame, housing pillar, standard
Ständerschwabbelmaschine pedestal buffer
starr positive
Starrfett consistent fat
Starrheit rigidity
Stärke gage, gauge, thickness
Staub dust
Staubabsauger dust extractor
Staubabsaugevorrichtung dust removing attachment
Staubabsaugung dust removal
staubdicht dust proof
Staubfänger dust collector
staubfrei dust proof
Staubschleifen d. h. mit pulverförmigen Mitteln unter Verwendung von Kluppen oder Scheiben aus Gußeisen, Stahl usw.) lapping
Stein brick, stone
Steinhalter, auswechselbarer refillable stoneholder
Stellmutter adjusting nut
steuern to control, to actuate
Steuerung control
Stichmaß gage, gauge
Stift pin, stud
Stirnfläche face
Stirnrad spur gear
Stirnradschleifmaschine spur gear grinding machine
Stirnseite (der Scheibe) grinding face
Stoß, geschliffener ground joint
stoßfrei free from shock
Stoßverbinder butt joint
Streichriemen abrasive strop, razor strop
Streichriemen-Paste razor strop dressing
Streuung, dichte close coating
Strom current
Stück, aus einem integral, solid
Stückform (Klumpen) lump form
Stücklohn piece rate
Stufe step
Stufenmotor adjustable speed motor
Stufenräderkasten feed gear box
Stufenscheibenantrieb cone pulley drive
. . . . stufig . . . . stepped
stumpfe Klinge dull blade
stumpf werden to grow tired, or dull
Support carriage

synthetisch synthetic(al)
Schaber scraping tool
Schabeverfahren scraping method
Schablone template, templet
Schafleder sheepskin
Schaft shank
Schale cup
schalten to control, to engage, to index
Schalter switch
scharf sharp
scharfe Schneide keen edge
scharf halten to keep keen
Scharten entfernen to take cut nicks
Schärfe edge, keeness, sharpness
schärfen to sharpen
Schärfen edging, sharpening
Schärfer sharpening stone
Schärfvorrichtung sharpening device
Scharte nick
Scheibe disc, disk, wheel, plate
Scheibe anschlagen to tap wheel
Scheibe, einfache glatte plain straight wheel
Scheibe- gegossene puddled wheel
Scheibe durch Klang prüfen to sound a wheel
Scheibe in Gummibindung rubber bonded wheel
Scheibe in Schellackbindung shellac bonded wheel
Scheibe in Silikatbindung silicate process wheel
Scheibe in Sonderform special shaped wheel
Scheibe mit abgerundetem Rand round faced wheel
Scheibe mit konischem Rand bevel face wheel
Scheibe zum Flächenschliff von Beton concrete surfacing wheel
Scheibenabdrehstein wheel dressing stick
Scheibenabnutzung wheel wear
Scheibenabziehwerkzeug wheel dresser
Scheibenauswuchtvorrichtung wheel balancing equipment
Scheibenbelag disc coating
Scheibenbruch wheel breakage
Scheibenersatz wheel replacement
Scheibenform wheel shape, shape of wheel
Scheibenform vordrehen to turn the shape
Scheibengefüge wheel structure
Scheibengeschwindigkeit wheel speed
Scheibengeschwindigkeit herabsetzen to slow down the wheel speed
Scheibengeschwindigkeit heraufsetzen to slow up the wheel speed
Scheibenkopf wheel head
Scheibenlebensdauer wheel live
Scheibenpressung wheel press
Scheibenschleifmaschine disc sanding machine
Scheibenrand wheel edge
Scheibenzusammenstellung disc assortment
Schellack shellac
Schellack-Bindung shellac bond
Scherenschleifer scissors grinder
Scherenschleifmaschine shear grinding machine, scissors grinder
scheuern to furbish
Scheuern scouring
Schicht, feine film
Schieblehre sliding caliber
Schienenschleifmaschine track-grinding machine
Schienenstoßschleifmaschine rail joint grinder
Schlag (einer Scheibe) excentricity
Schlagende Scheibe watbly wheel true pumping wheel
schlagfreie Scheibe true grinding wheel
Schlämmkreide spanish white, whitening
Schlangenmarken snake marks
Schleifahle hosting none
Schleifanstalt grinding mill, grinding plant
Schleifapparat grinding apparatus
Schleifarbeit grinding work
Schleifart type of grinding
Schleifartikel abrasive article
Schleifauflage grinding rest
Schleifausrüstung grinding equipment
Schleifband abrasive belt, belt, grinding belt, grinder belt, sanding belt
Schleifbank grinding bench, grinding lathe
Schleifblöcke zum Schleifen von Walzen roll scouring blocks
Schleifbock wheel stand, wheel slide, floor stand grinder
Schleifbürste polishing brush
Schleif-Drehplatte wheel stand plate

Schleif-Schieber wheel stand slide
Schleifdorn grinding arbour, lap plug
Schleifdruck grinding pressure
Schleifdurchmesser diameter to be ground
schleifen to grind, to sharpen, to whet
schleifen, mit der Bedeutung des Bremsens to drag
Schleifen grinding
Schleifen auf Einzylinder-Schleifmaschine single drum sanding
Schleifen auf Mehrzylinder-Schleifmaschine multiple drum sanding
Schleifen in der Werkzeugmacherei tool room grinding
Schleifen mit feinerer Körnung (2. Arbeitsgang beim Glasschliff) fining-up operation
Schleifen von Hand offhand-grinding
schleifend abradant
Schleifer grinder
Schleiferei grinding department, grindery, grinding mill
Schleiffachmann grinding expert
Schleiffeile abrasive file
Schleiffläche grinding surface
Schleifglas grinding glass
Schleifhobel abrasive plane
Schleifingenieur abrasive engineer
Schleifkante grinding edge
Schleifkissen grinding pad
Schleifkopf grinding head
Schleifkorn grinding grain, abrasive grain
Schleifkunst grinding art
Schleiflage grinding position
Schleiflänge length to be ground
Schleiflehre grinding gauge
Schleifleinen abrasive cloth
Schleifmarke grinding mark
Schleifmaschine grinding machine, grinder
Schleifmaschine, Abgrat- snagging grinding machine
Schleifmaschine, Abstech- cut-off grinding machine
Schleifmaschine, automatische automatic grinding machine
Schleifmaschine, hängende over head grinding machine
Schleifmaschine, für schwere Stücke heavy duty grinder
Schleifmaschine mit biegsamer Welle flexible shaft grinder
Schleifmaschine mit Fußantrieb foot grinder
Schleifmaschine mit Führungsstange guide-bar grinder
Schleifmaschine, halbautomatische semi-automatic grinding machine
Schleifmaschine mit Kraftantrieb power grinder
Schleifmaschine spitzenlose centerless grinding machine
Schleifmaschinenlünette grinder rest
Schleifmaschinenspindelstock grinding machine head stock
Schleifmehl emery flour
Schleifmittel, abradant, abrasive, grinding material
Schleifmittelbelag, letzter final coat of abrasive
Schleifmittel, künstliches manufactured abrasive
Schleifmittel, natürliches natural abrasive
Schleifmühle grinding mill
Schleifpapier abrasive paper
Schleifpapier mit offener Streuung open coated abrasive paper
Schleifpaste grinding compound, finishing compound
Schleifplatte grinding plate
Schleifplatz grinding stand
Schleifpolieren polish grinding
Schleifpraxis grinding practice
Schleifpulver grinding powder
Schleif-Querkratzer abrasive cross marks
Schleifrad grinding wheel, abrasive wheel
Schleifraum grinding room
Schleifraum-Abdrehvorrichtung wheel shaping device
Schleifring grinding cylinder, slip ring
Schleifrohstoff crude abrasive
Schleifsachverständiger grinding expert
Schleifsand abrasive sand
Schleifscheibe abrasive disc, abrasive wheel, grinding wheel
Schleifscheibe aus Schleifpapier oder -leinen sander disc, pin wheel (shoe)
Schleifscheibe, fliegende overhung grinding wheel
Schleifscheibe, schlecht schneidende lazy wheel
Schleifscheiben-Abdrehvorrichtung wheel dressing device, grinding wheel dresser

Schleifscheiben-Abdrehwerkzeug grinding wheel dresser
Schleifscheiben-Abrichter truing device
Schleifscheiben-Abrichtvorrichtung wheel truing attachment
Schleifscheiben-Antriebsscheibe wheel driving pulley
Schleischeiben-Dorn grinding wheel arboor spindle
Schleifscheiben-Schlitten grinding wheel slide
Schleifscheiben-Schutzhaube wheel guard
Schleifscheiben-Schutzkappe protector for grinding wheel
Schleifscheiben-Spindel grinding wheel spindle
Schleifschlitten wheel carriage
Schleifschmant grindstone swarf, grinding swarf
Schleifsegment abrasive segment
Schleifspezialität abrasive speciality
Schleifspindellager spindle box
Schleifspitze abrasive point
Schleifspitze, zahnärztliche dental point
Schleifstahl grinding bench
Schleifstaub emery dust, grindstone dust
Schleifstein grindstone, abrasive stone
Schleifsteinpulver wheel swarf
Schleifsteintrog grindstone box, grindstone through
Schleifsteinwelle grind wheel spindle
Schleifstellung grinding position
Schleifsupport grinding rest
Schleiftasse cupped wheel
Schleif- und Poliervorrichtung grinding and buffing attachment
Schleifteilkopf dividing head for grinding machine
Schleifverfahren auf der spitzenl. Rundschl.-Masch. straight-in and straight-trough
Schleifvorgang grinding operation
Schleifvorrichtung grinding attachment
Schleifwalze drum wheel, grinding roll, drum sander
Schleifwasser sanding liquid
Schleifwelle grinding wheel spindle
Schleifwerkstatt grinding shop
Schleifwinkel grinding angle
Schleifzeit grinding time
Schleifwerkzeug grinding tool
Schleifzubehör grinding equipment
Schleifzugabe grinding allowance
Schleifzwecke, allgemeine general grinding
Schleifzyinder grinding cylinder
schlichten to finish, to planish, to smooth
Schlichten finishing cut
schlichtschleifen smoth polish
Schlichtschnitt smoth polish
Schlichtstahl smothing tool
Schlitten slide
Schlitten, drehbar swivel rest
Schlitz slot
Schlitzfräser slot cutter
Schmelzofen furnace
Schmiedestein blacksmith's stone
Schmiege bevel
Schmiereinrichtung lubricator
schmieren to grease, to lubricate
Schmiermittel lubricant
Schmieröl lubricating oil
Schmierung lubrication
Schmirgel emery
Schmirgelband emery belt
Schmirgelbindemittel putty of emery
Schmirgelblock emery brick
Schmirgeleisen lap
Schmirgelfeile emery file
Schmirgelholz emery stick
Schmirgelkuchen emery cake
Schmirgelleinen emery cloth
Schmirgelmaschine emery wheel grinder
schmirgeln to emery, to rub with emery to grind with emery
Schmirgelpapier emery paper
Schmirgelpulver emery dust, emery flour, emery powder
Schmirgelscheibe emery wheel
Schmirgelscheiben-Abdrehverkzeug emery wheel dresser
Schmirgelschleifmaschine emery grinding machine
Schmirgelstaub emery dust
Schmirgelstein emery brick, emery stone
Schmirgelzylinder emery cylinder
Schnabel jaw
Schnecke spiral, worm
Schneckenschleifmaschine worm grinding machine
Schneide edge, lip of a tool
Schneide, nicht kratzende non-pulling edge
Schneide, scharfe keen edge
schneiden to cut

schneiden (in der Steinindustrie) to cope
Schneiden cutting, coping
Schneidfähigkeit cutting ability, c. efficiency, cutting quality
Schneidhärte cutting hardness
Schneidkante cutting edge, cutting lip
Schneidkraft cutting power
Schneidplättchen tool bit
Schneidscheibe cutting-off wheel
Schneidvorgang cutting action
Schneidwerkzeug cutting tool
Schneidwinkel angle of keenness
Schnellgang quick motion, rapid motion
schnellaufende Scheibe high speed wheel
Schnitt, sauberer sheer cut
Schnittbewegung cutting motion
Schnittgeschwindigkeit speed of cut, cutting speed
Schnitttiefe depth of cut
Schnittwinkel cutting angle, angle of keenness
schräg oblique
Schrägmaß bevel-square
Schrägschliff der Zähne sharpening the teeth to alternating angles
Schraube bolt
Schraublehre caliber rule
Schraubenkopf-Polierbank screw polishing lathe
Schraubenkopf-Poliermaschine screw polishing machine
Schraubenrad helical gear
Schraubenrad-Schleifmaschine spiral gear grinder
Schraubzwinge C-clamp
Schreibspitzen für Graveurstichel pencil points for engravers' chuck
schruppen to rough
Schruppen roughing, roughing cut
Schruppschleifen to rough, to snag
Schublehre slide calipers
Schuhmacher-Abziehstein shoemaker's sharpening stone
Schuhmacher-Streichriemen shoemaker's strop
Schutzbrille goggles
Schutzbügel guard
Schutzflanschen protecting flanges
Schutzhaube protecting cap, — cover, — guard, — hood, protection hood, safety hood
Schutzkappe hood
Schutzkappe für Schleifscheibe protector for grinding wheel
Schutzmaske face guard
Schutzvorrichtung safeguard, safety device
Schwabbelmaschine buffer, buffing lathe
Schwabbeln to buff
Schwabbeln buffing
Schwabbelscheibe buff, buff wheel, buffing wheel, ray wheel, glazer
Schwalbenschwanz dovetail
Schwalbenschwanz, erhöhter raised dovetail
Schwalbenschwanz, versenkter countersunk dovetail
Schwalbenschwanzförmige Verbindungsstelle (für Schleifpapier) splice-dovetail
Schweißnaht welded seam
schwenkbar swivable
schwenken to swing
Schwenkplatte swivel
schwere Arbeitsmaschine heavy duty machine
schwere Leinenscheibe heavy duty disc
Schwerpunkt center of gravity
schwingen to swing, to swivel
Schwingrahmenschleifmaschine swing frame grinder

Talg tallow
Talkspat magnesite
Taschenmesser-Abziehstein pocket hone
Tassenscheibe cup wheel, saucer wheel
Tassenscheibe, zweiseitige double cup wheel
Taster caliper, feeler
teilen to index
Teilscheibe index dial
Tellerrad ring gear
Tellerscheibe dish wheel
tempern to anneal
Tiefe der Aussparung depth of recess
Tiefenschleifen plunge cut grinding
Tiefenschliff-Verfahren in-feed grinding method
Tisch table, plate
Tisch-Bewegung table traverse
Tischfeststellschraube clamping bolt
Tischfläche table area
Tisch-Flächenschleifmaschine table surface grinder
Tischlauf-Steuerung table travel control

Tischlerstein jointer stone
Tischschlitten sliding table
Tisch-Schwenkplatte swivel table
Tisch-Schwenkschraube swivel table knob
Tischskala table scale
Tischumkehranschlag table reversing dog
Tischvorschub von Hand table-handfeed
toter Gang backlash, end play, lost motion
Toleranz limit, tolerance
Toleranzlehre limit gauge
Ton, basischer aluminous clay
Tonerde alumina
Topfscheibe pot wheel, tub-, cup-
Topfscheibe, gerade straight cup wheel
Topfscheibe, sich nach außen erweiternde flaring cup wheel
Transmissionswelle line shaft
transportable Schleifmaschine portable grinder
Transportscheibe feed wheel, regulating wheel
Trebel-Abdrehwerkzeug trebel wheel dresser
trennen to cut, so sever
Trennscheibe cutting-off wheel
Tripel tripoli powder, rotten stone
Trockenraum dryroom
trockenpolieren to pol'sh dry, dry polish
trockenschleifen to grind dry; dry grinding
Trockenschliff dry grinding
trocknen to bake
Trog trough
Trommel drum, barrel
trommeln tumbling
Trommelschleifmaschine drum grinding machine
Tuchscheibe buff wheel

Überhang overhang
übersetzen, ins langsame to gear down
übersetzen, ins Schnelle to gear up
Übertragung transmission
Überweg overtravel
Überzug, feiner film
Umdrehung revolution
Umdrehungszähler speed indicator, tachometer
Umfang periphery
Umfangsgeschwindigkeit circumferential speed, peripheral speed
Umfangsgeschwindigkeit in Fuß/min. s. f. p. m. S. F. P. M. surface feed per minute
umlaufen to circulate
Umlaufsgeschwindigkeit rotary speed
Umlaufszahl, minutliche revolutions per minute, r. p. m.
umlegen, einen Schalthebel to throw over a lever
Umriß outline
Umsteuermotor reversing motor
unappretiert unfinished
unausgebleit unbushed
unbearbeitet crude, unf'nished
unbearbeitete Feile unfinished stick
unbelegt uncoated
Ungleichheit unevenness
Ungleichmäßigkeit unevenness
ungleichmäßige Abnutzung uneven wear
Universal-Schleifmaschine universal grinder
Universal-Spannfutter universal chuck
Universal-Werkzeugschleifmaschine universal tool grinder
unmittelbarer Motorantrieb direct motor drive
unrund untrue, out of roundness, out true
unrund werden to become untrue, to get out of true
Untergestell frame, base
Unterlage (für Schleifbelag bei Schleifpapier und -leinen) backing
desgl., schwere Qualität aus Leinen backing heavy cloth
Unterlegscheibe washer
Unterteil base, bed
ununterbrochen continuous
unvollendet unfinished

Ventil valve
Ventil-Einschleifmaschine valve refacer
Ventil-Schleifpaste valve grinding compound
Verbindung joint
Verbindungsstelle splice
verbrennen to burn
Verfahren process
verfeinern to finish, fine grinding
Vergleichsstück master block

Verharzen gumming
Verhältnis, rate, ration
Verjüngung taper
verkitten to cement
verleihen, einem Werkzeug eine scharfe Schneide to impert a cutting edge
verriegeln to interlock with, to lock
verrippt webbed
Versagen break down
versenken to countersink
versenkter Schwalbenschwanz countersunk dovetail
versetzt off-set
Verschleiß wear
verschmieren to clog, to fill
Verschmieren (einer Schleifscheibe), glazing, loading
verschmierte Scheibe loaded wheel
verschrauben to bolt
versintern to vitrify
Vertiefen slotting
Vertiefung slotting recess
Verzahnung toothing
Verzierschleifen decorative grinding
Vierkantfeile square stick
vollkommen genau dead true
vollsetzen (einer Scheibe mit Spänen) to clog
von Hand by hand, off hand
Vordreherei (für gegossene Rohscheiben) shaping shop
Vorgelege countershaft, back gear
Vorläppen rough lapping
Vorpolieren first polishing, pre-polish
Vorrichtung appliance, attachment, device fixture, equipment
vorschleifen to grind first, — — preparatory, — — grind rough
Vorschleifmasse pregrinding paste
Vorschub feed
Vorschubauslösung feed tripping
Vorschubbereich feed range
Vorschubbewegung feed motion
Vorschubbewegung im Winkel angular feed
Vorschubbewegung der Schleifscheibe traverse of the wheel
Vorschubgeschwindigkeit traverse speed
Vorschubwechsel feed change
Vorsprung (an einer Scheibe) offset
Vorsteckscheibe C-washer

waagerecht level
Wandstärke (einer Topfscheibe) thickness rim
Walroßleder walrus leather
Walze drum, roll
walzen to mill
Walzenschleifen grinding rolls, roll-grinding
Walzenschleifmaschine roll grinding machine, drum sander
wasserdicht waterproof
wasserdichter Abschluß water seal
wasserdichtes Schleifpapier waterproof grinding paper
Wasserdüse water nozzle
wasserfest waterproof
Wasserstein water hone, whet stone
Wasserzuführungseinrichtung water attachment
Wasserzuführungsrohr water piping
Wälzlager anti-fricition bearing
Wechselstrom A. C. (alternating current)
Wechselrad change gear
wegnehmen to take off
weiche Bindung soft bond
Welle shaft, spindle
Welle, biegsame flexible shaft
Werfen (der Scheibe) warping
Werkbank-Schleifmaschine bench grinding machine
Werkstatt shop
Werkstatt-Mikroskop shop-microscope
Werkstück work-piece
Werkstück-Auflage work support
Werkstück-Antriebsscheibe live spindle driving pulley
Werkstück-Durchmesser diameter of work
Werkstück-Geschwindigkeit work speed
Werkstück-Halter work holder
Werkstück-Stütze work support
Werkstück-Zentriervorrichtung work locating device
Werkzeug tool
Werkzeugausgabe tool room
Werkzeuggestell tool rack
Werkzeughalter tool holder, tool block
Werkzeughalter, gekröpfter angle tool holder
Werkzeuglager tool room
Werkzeugmacherei tool room
Werkzeugmarken tool marks

Werkzeug-Naßschleifmaschine wet tool grinder

Werkzeugschleiferei tool grinding department

Werkzeugschleifer tool dresser

Werkzeugschleifmaschine tool grinder

Werkzeugschlitten slide rest, carriage

wetzen to whet

wieder ausbleien to rebush

wieder herrichten to furbish up

Wiederpolieren re-polisn

Wiener Kalk Vienne lime

Winkel angle, square, rake

Wnkel, Freischnitt clearance angle

Winkel, rechter acute angle

Winkel, Span- rake angle

Winkel-Abgratmaschine angle dressing machine

Winkelfräser angular cutter

Winkelgeschwindigkeit angular feed

Winkelhebel bellcrank

Winkeltisch knee

Winkelzahnrad angular wheel

winklich angular

Wirkungsgrad efficiency

wirtschaftlich ecenomical

Wolfram tungsten

Wolframcarbid tungsten carbide

Wolframcarbid-Werkzeug tungsten carbide tool

Wulst collar

Zahnauflage tooth rest

zahnärztliche Schleifspitze dental point

Zahnhöhe depth of tooth

Zahnrad gear

Zahnrad mit Innenverzahnung internal gear

Zahnradpumpe gear pump

Zahnstange rack

zähe Bindung tough bond

Zähigkeit toughness

Zangenfutter draw back chuck

Zapfen pin, plug

Zeiger index

zentrieren to locate

Zentriermaschine centering machine

Zentrierung location

Zentriervorrichtung locating device

Zentrischschleifen spot grinding

zerbrechlich brittle

zerkleinern to crush

Zerkleinerung crushing

Zerkleinerungsanlage crushing plant

Zerkleinerungsraum crushing room

Zerreißfestigkeit tensible strength

zerspanen to chip

zerspringen der Scheibe flying off of the wheel

Zettel label

Ziehschleifen honing

zieren decorate

Zifferblatt dial

Zinnasche, Zinnoxyd putty powder

Zittermarke chattermark

Zittern (des Werkzeugs) chattering

Zollstock rule

Zubehör accessory, appliance

Zuführungsrohr feed tube

Zuführungswalze feed roller

Zugabe addition, allowance

zugesetzte Scheibe loaded wheel

Zugfestigkeit tensile strength

zunehmen to increase

Zusammenbau assembly

zusammenbauen to assemble

Zusammensetzung composition

Zuschlag addition

zusetzen to fill

Zustellung feed, adjustment

zwangläufig positive

zwängen, eine Scheibe auf die Spindel to crowd a wheel on the spindle

zweiseitig ausgesparte Scheibe double recess wheel

zweiseitige Tassenscheibe double cup wheel

Zwillings . . . twin . . .

Zwischenlage washer

zylindrisch cylindrical, straight

Zylinderscheibe cylinder wheel

Zylinderschleifen cylindrical grinding

Zylinderschleifmaschine cylindrical grinding machine

Zylinder-Stichmaß caliber gauge

Dictionary

of

Grinding and Polishing Terms

Part II

English-German

Edited by

Senator h. c. B. Kleinschmidt

Preface

In every industry many idioms or colloquial terms are used that are not included in any ordinary dictionary. The work oft the translator is further complicated by the fact that there is considerable variation in many of these special terms as applied to a given process, material, tool etc.; consequently, the practical value of a technical dictionary depends largely, aside from its general scope, upon the number of important colloquial terms it contains and upon the accuracy of their definitions.

The purpose of this dictionary is to serve the abrasive engineer and the user of the grinding and polishing materials. It contains all the grinding and polishing terms, which on account of their novelty have not yet been taken up by other technical dictionaries.

Polishing

Polishing is that operation where wheels of any kind such as felt, canvas, leather, walrus, wood leather, faced or stiched cotton buffs are "set up" on their faces with grain abrasives in various grits. The Automatic Buffing machine is used to a great extent for polishing, but more largely for buffing.

Buffing

Buffing is divided into two separate sub-operations: a) „cutting down" and b) "coloring up".

Cutting down is that operation of finishing the surface of an article, taking out imperfections left by rolling, stamping, or previous polishing operations. The cutting down operation approximates a high finish.

Coloring up is that operation that brings the surface to a high luster, removing the grease, dirt and finger marks left by the previous operation of cutting down. Coloring up is also that operation that follows plating to give the plated part a luster.

abradant, reibend, schleifend, Schleifmittel, Schleifpulver

abrade, to abschleifen, abschaben, abreiben

abrasion, Abschleifung, Abschürfung, Abschabung

abrasive, to abschleifen

abrasive, Schleifmittel, Schleifmaterial

abrasive article, Schleifartikel

abrasive-belt, Schleifband

abrasive-cloth, Schleifleinen

abrasive cross marks, Schleif-Querkratzer

abrasive-disc, Schleifscheibe

abrasive-engineer, Schleifingenieur

abrasive-file, Schleiffeile

abrasive-grain, Schleifkorn

abrasive-material, Schleifmaterial

abrasive-paper, Schleifpapier

abrasive-planer, Schleifhobel

abrasive-point, Schleifspitze

abrasive-powder, Schleifpulver,

abrasive-sand, Schleifsand

abrasive-segment, Schleifsegment

abrasive-speciality, Schleifspezialität

abrasive-stone, Schleifstein

abrasive-strop, Streichriemen

abrasive-wheel, Schleifrad, Schleifscheibe

abrasive-wheel department, Schleifscheiben-Abteilung

abrasive-wheel segment, Schleifscheiben-Segment

A. C. (alternating current), Wechselstrom

accessory, Zubehör

accurate grinding, Präzisionsschleifen

acicular, nadelkristallförmig

acute angle, rechter Winkel

adapter, Paßstück

addition, Zuschlag, Zugabe

adjust, to eichen, einpassen, richten

adjustable, einstellbar, verstellbar

adjustment, Einstellung, Einrichtung, Zustellung

alignment, Ausrichtung

alligator cracks, feine Risse (z. B. in Tiegeln)

allowance, grinding - Schleifzugabe

alloy, Legierung

alloy-steel, legierter Stahl

alternate motion, hin- und hergehende Bewegung

alumina, Tonerde

aluminous abrasive, tonerdehaltiges Schleifmittel, Elektrokorund, Korund-Schleifmaterial

aluminous clay, basischer Ton (mit geringem Silikatgehalt)

aluminum, Aluminium

aluminum oxide, Aluminium-Oxyd Elektrokorund

amount of stock, Materialmenge

amount of taper, Betrag der Konizität, Kegelneigung

angle, Winkel, Winkeleisen

angle dressing machine, Winkeleisenabgratmaschine

angle, clearance, Freischnittwinkel

angle, grinding - Schleifwinkel

angle, rake - Spanwinkel

angle, of keenness, Schneidwinkel

angle tool holder, gekröpfter Werkzeughalter.

angular, winklig

angular cutter, Winkelfräser

angular feed, Winkelgeschwindigkeit

angular thread, Spitzgewinde

angular wheel, Kegelrad, Winkelzahnrad

anneal, to glühen, tempern

appliance, Vorrichtung, Zubehör

arbor, arbour, Dorn, Spindel

arbour, grinding - Schleifdorn

arbour, grinding wheel Schleifscheibenspindel

arbour hole, Spindelloch
arc, Bogen
area, Fläche
area of grinding contact, Berührungsfläche zwisch. Schleifscheibe u. Werkstück
arkansas stone, Arkansas-Stein
artificial abrasive, künstliches Schleifmittel
artificial corundum, Kunstkorund
artificial resin, Kunstharz
assemble, to montieren, zusammenbauen
assembly, Montage, Zusammenbau
attachment, Vorrichtung (für Zusatzarbeiten)
attack, to ätzen
auger, Holzbohrer
attack-polish, to ätzpolieren
automatic (ally), automatisch
auto-stone, Autostein
average output, mittlere Leistung
axe stone, Axtstein

back off, to hinterschleifen
back rest, Satzstock
backing, Unterlage für Schleifbelag b. Schleifpapier und -leinen
backing, heavy cloth - desgl. aus schwerem Leinen
backlash, toter Gang
bake, to trocknen
bakelite, Bakelit
balance, to auswuchten, ausbalancieren
balancing, Ausbalancieren
ball grinding machine, Kugelschleifmaschine
ball wheel, Kugelscheibe
barb, Grat
barrel, Trommel
barrel, tumbler - Putztrommel
base, Unterteil, Maschinenbett
Bauxite, Bauxit
bearing, Lager, Führung
bearing, anti-friction - Wälzlager
bearing, ball - Kugellager
bearing, plain, glattes Lager
bearing, roller - Rollenlager
bearing sizing, Polierrollen von Bohrungen
bearing, taper roller - Kegelrollenlager
become untrue, to unrund werden
bed, Maschinenbett
bed water guard, Wasserschild zum Schutze des Maschinenbettes
belt, Band, Riemen
belt, emery - Schmirgelband, Schleifband
belt driven, riemengetrieben
belt, endless endloses Band
belt-grinder, Bandschleifmaschine
belt grinding machine, Bandschleifmaschine
belt-polisher, Polierriemen
belt polishing machine, Bandpoliermaschine
belt, sanding - Schleifband
bench, Maschinenbett, Bank
bench, grinding - Schleifbank
bench grinding machine, Werkbankschleifmaschine
bench stone, Bankstein
bevel, Schmiege, Facette
bevel to facettieren
beveled face, abgeschrägte Fläche
bevel face wheel, Scheibe mit konischem Rand
bevel gear grinding machine, Kegelradschleifmaschine
bevel grinding attachment, Konischschleifvorrichtung
bevelling the front edge of teeth, Schrägschliff der Zähne
bevel-square, Schrägmass
binding material, Bindemittel
bit, tool - Schneidplättchen
blacksmiths' stone, Schmiedestein
blade, Klinge, Messer
blade, inserted - eingesetztes Messer
blank, Rohstück, unbearbeitetes Werkstück
blast, sand - Sandstrahlgebläse
block, Block
block, precision gauge - Endmass
block, tool - Werkzeughalter
blood-stone, Blutstein, Roteisenstein, Polierstein.
blotter, Pappflansch
blotting paper, Löschpapier
blow, to blasen
bolt, Schraube
bolt, to verschrauben
bond, Bindung (bei Schleifscheiben)
bonding agent, Bindemittel
bonding process, Bindungsart
bore, Bohrung
boring, Bohrung

box board lap, Holzpolierscheibe
box, feed gear - Stufenräderkasten
box, guide, Führungsbuchse
box, speed change, Getrieberäderkasten
bracket, Konsole, Sockel, Lager
brake drum drinding machine, Bremstrommelschleifmaschine
brake drum grinder, Bremstrommelschleifmaschine
brass, Messing
breakage, Bruch
breakdown, Bruch, Versagen
brick, Stein
brighten, to, glätten, polieren
brilliant polish, Hochglanzpolitur
brittle, spröde, brüchig, glashart, zerbrechlich
brittleness, Sprödigkeit, Brüchigkeit, Glashärte
broach, Räumnadel
brush, polishing - Polierbürste
brush, wire - Drahtbürste
buff, to glanzschleifen, feinpolieren, hochglanzpolieren, schwabbeln
buff, Polierscheibe, Schwabbelscheibe,
buff leather, Polierleder
buff stick, Polierscheiben-Abdrehwerkzeug
buff wheel, Schwabbelscheibe, Tuchscheibe
buffer, Schwabbelmaschine
buffing, Schwabbeln, Feinpolieren (Vorfeinpolieren und Fertigfeinpolieren)
buffing cone, Polierdorn
bufing lathe, Schwabbelmaschine
buffing machine, Poliermaschine
buffing powder, Polierpulver
buffing wheel, Polierscheibe
bumping wheel, schlagende Scheibe
burn, to brennen, verbrennen
burn the stock, to, das Material verbrennen
burnish, to glanzschleifen, glätten, hochglanzpolieren
burnisher, Polierstahl, Glättstahl
burnisher, scraper, Polierschaber
burnishing, Polieren, Glätten, Glänzen, Polierrollen von Zahnrädern
burnishing machine, Drückbank, Poliermaschine
burnishing stone, Polierstein, Blutstein
burnishing tool, Polierwerkzeug
burr, Grat, Rohrgrat, Preßgrat
burr, to abgraten, entgraten
burring, Abgraten
burring machine, Abgratmaschine
bursting of the wheel, Bersten d. Scheibe
bush, Buchse
bushed, ausgebleit
bushing, Buchse, Lagerbuchse
butt joint, Stoßverbindung
button sleeker, Polierlöffel
by hand, von Hand
by-product, Nebenprodukt

C-clamp, Schraubzwinge
C-washer, Vorsteckscheibe mit C-Schlitz
cage, ball - Kugellagerkäfig
caliber, caliper, Kaliber
caliper gauge, Zylinderstichmaß
caliper gauge grinder, Lehrenschleifmaschine
caliper rule, Schraublehre, Kalibermaßstab
calipers, Taster, Tasterzirkel
calipers, micrometer, Mikrometerlehre
calipers, slide, Schieblehre
cam, Nocken, Kurvenscheibe, Excenter
cam-actuated (or controlled), durch Kurve gesteuert
cam grinding machine, Nockenschleifmaschine, Nockenwellenschleifmaschine
cam shaft grinding attachment, Nockenwellenschleifeinrichtung
cambering attachment, Balligschleifeinrichtung
canvas, Kanvas (Segeltuch)
capacity, Leistung, Leistungsvermögen, Fassungsvermögen
caput mortuum, Englisch Rot
car-wheel grinder, Radsatz-Schleifmaschine
carbide of silicon, Silizium-Karbid
card-clothing grinder, Kratzenbelag-Schleifmaschine
card-cylinder-grinder, Kardenzylinder-Schleifmaschine
carriage, Support, Werkzeugschlitten
carriage, grinding, Bettschlitten
carvers' slip, Fassonabziehstein für Bildhauer
cement, to leimen, verkitten
cementing in of the grinding wheel, Einkitten der Schleifscheibe
center, dead - feststehende Spitze

center grinder, Spitzeschleifapparat, Spitzenschleifmaschine
center, grinding attachment, Körnerspitzenschleifapparat
center grinding machine, Spitzenschleifmaschine
center, live - umlaufende Spitze
center of gravity, Schwerpunkt
center, off - außerachsig
center point, Körner, Reitstockspitze
center, tailstock - Reitstockspitze
centering machine, Zentriermaschine
centerless grinding, spitzenlose Schleifen
centerless grinding machine, spitzenlose Rundschleifmaschine
centers, height of Spitzenhöhe
ceramic bond, keramische Bindung
change feed, Vorschubwechsel
change over, to, umschalten
charge to - a wheel or lap, etwas auf eine Scheibe oder ein Poliereisen auftragen.
chatter, Erschütterung, Rattern
chatter mark, Rattermarke, Zittermarke
chattering. Zittern (des Werkzeuges)
check, Riß, Haariß (im Stahl)
check the speed, to, die Geschwindigkeit einhalten
chemical composition, chemische Zusammensetzung
chip, Span
chip clearance, Spanbeseitigung
chip formation, Spanbildung,
chip, to, zerspanen, meißeln
chipping, Spanbildung
chrome oxide, Chromoxyd
chuck, Spannfutter, Bohrfutter
chuck, draw back, Zangenfutter
chuck, draw in, Spannpatrone
chuck, magnetic - magnetisches Spannfutter
chuck, pneumatic - Druckluft-Spannfutter
chuck, split - Spannpatrone
chuck, to einspannen
circular grinder, Rundschleifmaschine
circular grinding, Rundschliff, Rundschleifen
circular grinding machine, Rundschleifmaschine
circular saw, Kreissäge
circular saw for cutting metal, Metallkreissäge
circulate, to umlaufen (im Kreise)
circumferential speed, Umfangsgeschwindigkeit
clamp, Klemmhebel, Schraubzwinge
clamp, to einspannen, aufspannen
clamping bolt, Tischfeststellschraube
clamping lever, Feststellhebel
clamping plate, Spannplatte
clean, to, reinigen, putzen
clean up, to säubern
cleaner, Putzer
cleaning drum, Putztrommel
cleanse, to, reinigen
clear polish, klarschleifen
clear ring, reiner Klang (Scheibenprobe)
clearance, Spiel, Spielraum, Hinterschliff
clearance of the drill, Hinterschliff des Bohrers
clearance-ground, hohlgeschliffen
clog to verschmieren, mit Spänen vollsetzen
close coating, dichte Streuung
cloth, Leinen
cloth cutters' stone, Abziehstein für Tuchschneidemesser
cloth disc, Leinenscheibe
cloth economy roll, Leinen-Sparrolle
cloth, emery - Schmirgelleinen
cloth in odd width rolls, Leinenrollen verschiedener Breite
cloth pad, Leinenkissen
clutch, Kupplung
coarse, to grind, grobschleifen
coarse-grained, grobkörnig
coarseness, Grobheit
coat of abrasive, Schleifmittelbelag
coat, to bestreuen, belegen, überziehen
coated abrasives, (Sammelbegriff für Schleifpapier und -Leinen)
coating, Belag, Streuung
coating, single, einfacher Belag
coating, double, doppelter Belag
coil, Rolle
coke, Koks
colcothar, Englisch Rot
collar, Bund, Wulst
combination abrasive paper, Kombinations-Schleifpapier
combination bench stone, Kombinations-Bankstein

combination hone, Kombinations-Abziehstein
common salt, Kochsalz
compound, grinding - Schleifpaste
concentric, Konzentrisch
concentricity, Rundlaufen
concrete surfacing wheel, Scheibe zum Flächenschliff von Beton
cone, Konus, Stufenscheibe
cone, buffing- Polierdorn
cone grinding, Kegelschleifen, Konischschleifen
cone grinding roll, Konusschleifwalze
cone, sander - Sandpapierkonus
cone shaped cup wheel, konische Tassenscheibe
connection, Schaltung
constant, gleichmäßig
contact, Eingriff
continuous, ununterbrochen, pausenlos
control, Steuerung
control, to, steuern, regulieren, schalten
cool grinding, Naßschliff
coolant, Kühlmittel
cooling medium, Kühlmittel
cooling water, Kühlwasser
cope, to schneiden, (Steinindustrie) einkerben
coping, Schneiden
copper burnishing wheel, Kupferpolierscheibe
corrosion resistant alloys, korrosionsbeständige Legierungen
corundum, Korund
cost of wheel upkeep, Kosten der Scheiben-Unterhaltung
cotton dolly, Baumwollscheibe, ungenäht
cotton mop, zum Polieren
counterlockwise, entgegengesetzt zu Uhrzeigersinn
countershaft, Vorgelege
countersink, to, versenken
countersunk dovetail, versenkter Schwalbenschwanz
counterweight, Gegengewicht
coupling, Kupplung
crack, Riß, Härteriß, Sprung
crack, to, springen, Risse bekommen, bersten
crank, Kurbel
crank, bell - Winkelhebel
crank shaft, Kurbelwelle
crocus, Polierrot
cross feed, Querbewegung, Quervorschub (des Schleifrades)
cross feed, power - Selbstgang, Planzug
cross feed pawl, Sperrklinke für den Quervorschub
crowd, to - a wheel on the spindle, eine Scheibe auf die Spindel zwängen
crown, Wölbung
crowned, ballig
crude, unbearbeitet
crude abrasive, Schleifrohstoff
crush, to, brechen, zerkleinern
crusher, Brecher, Brechwalzwerk, Hartzerkleinerungsmaschine
crushing, Zerkleinerung
crushing plant, Zerkleinerungsanlage
crushing room, Zerkleinerungsraum
crystalline, kristallinisch
crystalline alumina, Korund, Aluminium-Oxyd
cupped wheel, Scheibe mit Aussparung
cup, Pfanne, Napf, Schale
cup, roller, bearing, Rollenlagerring
cup wheel, Tassenscheibe, Topfscheibe
cup wheel, cone shaped - konische Tassenscheibe
curl, to, rollen, biegen
current, Strom
current, alternating - Wechselstrom
current, direct Gleichstrom
curve, Kurve
curved, gekrümmt
cut, to schneiden, trennen
cut eleanly, to freischneiden
cut, depht of Schnittiefe
cut down, to feinpolieren
cut facets, to facettieren
cut, finishing - Schlichtschnitt, Feinschnitt
cut freely, to freischneiden
cut in and out, to ein -und ausschalten (Strom)
cut-off wheel, Schneidscheibe, Trennscheibe
cut, sheer sauberer Schnitt
cut to size, to auf Maß abdrehen
cutlers' stone, Messerschmiedewaren-Abziehstein
cutter, Abdrehrädchen, Fräser, Messer
cutter grinder, Fräserschleifmaschine

cutting ability, Schneidfähigkeit
cutting action, Schneidvorgang
cutting angle, Schnittwinkel
cutting edge, Schneidkante
cutting efficiency, Schneidfähigkeit
cutting hardness, Schneidhärte
cutting lip, Schneidkante
cutting motion, Schnittbewegung
cutting-off machine, Abstechmaschine
cutting-off tool, Abstechwerkzeug
cutting power, Schneidkraft
cutting pressure, Schneiddruck
cutting quality, Schneidfähigkeit
cutting tool, Schneidwerkzeug
cylinder grinding machine, Zylinderschleifmaschine
cylinder, grinding - Schleifring
cylinder wheel, Zylinderscheibe
cylindrical grinder, Rundschleifmaschine
cylindrical grinding, Rundschliff
cylindrical grinding attachment, Rundschleifapparat

D. C. (direct current), Gleichstrom
dead center, feststehende Spitze
dead center pulley, Antriebsscheibe für tote Spitze
dead smooth, spiegelglatt
dead true, vollkommen genau
decorate, zieren
decorative grinding, verzierschleifen
degree of accuracy, Genauigkeitsgrad
degree of hardness, Härtegrad
dense structure, dichtes Gefüge
dental point, zahnärztliche Schleifspitze
dental wheel, zahnärztliche Schleifscheibe
department, grinding - Schleiferei
department, tool grinding - Werkzeugschleiferei
depth of cut, Schnittiefe
depth of recess, Tiefe der Aussparung
depth of thread, Gangtiefe
depth of tooth, Zahnhöhe
Derbyshire spar, Flußspat
device, Vorrichtung, Mechanismus
deviation, Ausschlag (der Meßuhr)
dial, (index-), Zifferblatt (Teilscheibe)
diameter, Durchmesser
diameter of work, Werkstückdurchmesser
diameter to be ground, Schleifdurchmesser
diamond, Diamant
diamond dressing, Abdrehen mit dem Diamanten
diamond dust, Diamantstaub
diamond lap, Diamantschleifwerkzeug
diamond tool, Diamantwerkzeug
diamond tool holder, Diamanthalter zum Abdrehen der Schleifscheibe
diatomaceous earth, Diatomeenerde, Kieselgur
dimension, Maß, Abmessung
direct motor drive, unmittelbarer Motorantrieb
diassemble, to auseinandernehmen
disc coating, Scheibenbelag
disc crank, Kurbelscheibe
disc excentric, Exzenterscheibe
disc friction, Reibscheibe
disc grinder, Flächenschleifmaschine
disc grinder press, Leimpresse für Leinenscheiben
disc multiple, Lamelle
disc sander, Sandpapierschleifmaschine
disc sander, Schleifscheibe aus Schleifpapier oder -leinen
disc sanding machine, Scheibenschleifmaschine
dish wheel, Tellerscheibe
distance between center, Spitzenentfernung
dog, Mitnehmer, Anschlag
dog stop, verstellbarer Anschlag
dog table, reversing - Tischumkehranschlag
double cup wheel, Doppeltopfscheibe, zweiseitige Tassenscheibe
double recess wheel, zweiseitig ausgesparte Scheibe
dovetail, Schwalbenschwanz
dress, to abdrehen, abrichten, bearbeiten
dresser, Abdrehapparat, Abziehapparat
dresser, emery wheel - Schmirgelscheibenabdrehwerkzeug
dresser, grinding wheel - Schleifscheibenabdrehwerkzeug
dresser, metcalf wheel - Abdrehvorrichtung (Scheibe)
dresser, tool - Werkzeugschleifer
dresser, Trebel wheel - Trebel-Abdrehwerkzeug
dresser, wheel - Scheiben-Abziehwerkzeug
dressing, Abdrehen

dressing-practice, Abricht-Praxis
dressing-time, Abricht-Zeit
dressing-tool, Abdrehwerkzeug
drill grinder, Bohrerschleifmaschine
drill grinding machine, Bohrerschleifmaschine
drill socket, Bohrfutter
drill twist, Spiralbohrer
drive, Antrieb
drive cone pulley, Antrieb durch Stufenscheibe
driven belt, Antrieb durch Riemen
driven motor, Antrieb durch Motor
driver, Mitnehmer
drum, Trommel, Walze
drum brake, Bremstrommel
drum grinding machine, Trommelschleifmaschine
drum sander, Walzenschleifmaschine
drum sanding, single - Schleifen auf Einzylinderschleifmaschine
drum sanding, multiple - Schleifen auf Mehrzylinderschleifmaschine
drum wheel, Schleifwalze
dry grinder, Trockenschleifmaschine
dry grinding, Trockenschliff
dry polish, trockenpolieren
dryroom, Trockenraum
ductility, Dehnbarkeit
dull, to stumpf werden
dull blade, stumpfe Klinge
dull finish, ungenügender Fertigschliff
dull grinding, mattieren, mattschleifen
duplicate a wheel, to genau die gleiche Scheibe anfertigen
duplicating work, Reihenfertigung
duplication, Austauschbarkeit
durability, Dauerhaftigkeit
dust, Staub
dust collector, Staubfänger
dust exhaustor, Staubabsauger
dust extractor, Staubabsauger
dust proof, staubfrei, staubdicht
dust removal, Staubabsaugung
dust removing attachment, Staubabsaugevorrichtung
dwell of a cam, Kurvenscheibenteil, der Stillstand des bewegl. Tellers bewirkt

eccentric, exzentrisch
eccentricity, Schlag (von Scheiben)
economical, wirtschaftlich
economy roll, Sparrolle
edge, Schneide, Kante, Schärfe
edge lenses, to Linsenränder schleifen
edging, Schärfen, Kantenschliff
efficiency, Wirkungsgrad, Leistungsfähigkeit
elastic, elastisch
elastic process, elastisches Verfahren
elastic wheel, elastische Scheibe
electric drive, elektrischer Antrieb
electric furnace, elektrischer Schmelzofen
electric grinder, elektrische Schleifmaschine
electro furnace plant, Elektro-Schmelzwerk
electromagnetic, elektromagnetisch
elektroplating, Galvanisieren
emery, Schmirgel
emery belt, Schleifband
electro furnace-plant, Elektro-Schmelzgelstein
emery cake, Schmirgelkuchen
emery cloth, Schmirgelleinen
emery cylinder, Schmirgelzylinder
emery dust, Schmirgelstaub, Schleifstaub
emery flour, Schmirgelpulver, Schleifmehl
emery grinding machine, Schmirgelschleifmaschine
emery paper, Schmirgelpapier
emery powder, Schmirgelpulver
emery, putty of —, Schmirgelbindemittel
emery stick, Schmirgelholz, Schmirgelfeile
emery stone, Schmirgelstein, Schmirgelscheibe
emery, to schmirgeln
emery wheel, Schmirgelscheibe
emery wheel grinder, Schmirgelmaschine
end-feed, Zapfen-Anschleifen im spitzenlosen Schleifverfahren
endless belt, endloses Band
end play, toter Gang
engage, to eingreifen, einrücken, schalten
engagement, Eingriff
Englisch red, Englisch Rot, Polierrot
engravers' pencil point, Graveurstichel
equipment, Ausrüstung, Vorrichtung
equipment, grinding - Schleifausrüstung
erect, to montieren
etch-polish, to ätzpolieren
even, gerade, eben
even grinding, ebenschleifen
exhaust, to absaugen

exhauster, Exhaustor
exhaustion, Absaugung

fabric for wheels, Gewebestoff für Polierscheiben
face, Rand, Stirnfläche, Vorderseite
face grinding, Planschleifen
face grinding wheel, Kronenschleifstein
face guard, Schutzmarke
face milling cutter-grinding attachment, Messerkopfschleifapparat
face plate, Planscheibe
face, to flachfräsen, plandrehen, fasen
faced, belegt
faced, single - einseitig belegt
faced, double - doppelseitig belegt
facing off, Glattmachen (Glasindustrie)
false center, feststehende Spitze
fasten, to befestigen
feed, Vorschub, Zustellung
feed, angular - Vorschubbewegung im Winkel
feed, automatic - Selbstgang-Vorschub
feed, automatic fine - Selbsttätige Feineinstellung
feed, cross - Planbewegung, Plangang
feed, longitudinal - Längsbewegung, Längsgang
feed, power - Selbstgang
feed wheel, Transportscheibe, Vorschubscheibe
feeler, Fühllehre, Taster, Fühler
feldspat, felspar, Feldspat
felt, Filz
felt-pad Filzkissen
felt polishing belt, Filzpolierband
felt polishing disc, Filzpolierscheibe
felt washer, Filzscheibe
felted polisher, befilztes Polierholz
ferreous, eisenhaltig
fettle, to abgraten, grobschleifen
fettling wheel, Abgratscheibe, Grobschliffscheibe
file, Feile
file, to feilen
file, emery - Schmirgelfeile
file, polishing - Polierfeile
fill, to zusetzten, verschmieren
fill up, to sich zusetzen
filler, Porenfüller (beim Polieren)
film, feine Schicht, feiner Ueberzug
film, oil - Oelhaut
final coat of abrasive, letzter Schleifmittelbelag
fine bevels, to Facetten feinschleifen
fine-grained, feinkörnig
fine-grinding, verfeinern
fineness, Feinheit, Feinkörnigkeit
fineness of finish, Güte der Fertigschliffs
fining-up operation, Schleifen mit feinerer Körnung (2. Arbeitsgang beim Glasschleifen
finish, Feinschliff, Fertigschliff
finish, to schlichten, fertigmachen, verfeinern
finish, commercial finieren, handelsüblicher Schliff
finish, grinding Fertigschliff, Nachschliff
finish lapping, Fertigläppen
finish, satin mattgeschliffen
finish, semi-, to mattieren, mattschleifen
finish ultra, Spiegel-Politur
finished, bearbeitet, appretiert
finishing cloth, Feinschleifleinen
finishing compound, Polierpaste
finishing cut, Schlichten
finishing paper, Feinschleifpapier
firestone, Feuerstein
fit, Passung, Sitz
fit, to passen, sitzen
fixture, Vorrichtung
flange, Flansch
flexible, biegsam, schmiegsam
flexible cloth, biegsames Leinen
flexible shaft, biegsame Welle
flexible shaft grinder, Schleifmaschine mit biegsamer Welle
flint, Flint, Flintstein
flint paper, Flintschleifpapier
floating wheel, Flatterscheibe
floor stand grinder, Schleifbock
floor surfacer, Fußbodenschleifmaschine
floor surfacing machine, Fußbodenflächenschleifmaschine
flush, bündig, abschneidend
flute, Rille, Hohlkehle, Nut
flute, to riefeln
fluted, straight and spiral- gerade und spiralig gerillt
fluting, Hohlkehle
fly-wheel, Schwungrad
flying-off of the wheel, Zerspringen der Scheibe

foot grinder, Schleifmaschine mit Fußantrieb
footstock, Reitstock
footstock center, Reitstockspitze
force acting against, Anpreßdruck gegen
form grinding, Formschliff, Profilschliff
formed wheel, Formscheibe
foundation, Fundament, Aufstellung
frame, Rahmen, Untergestell
free cutting, Freischneiden
free from shock, stoßfrei
free hand grinding, Freihandschleifen
friction, Reibung
frost, to mattschleifen
furbish, to polieren, polierschleifen, scheuern, putzen
furbisher, Polierer, Putzer, Glätter
furbish up, to aufputzen, wieder herrichten, neuen Glanz geben
furnace, Schmelzofen
furrow, to auskehlen (Steinindustrie)
furrowing, Auskehlen

gage, (= gauge), to messen, anzeigen
gage, Stärke, Dicke, Lehre, Stichmaß, Meßvorrichtung
gage, twist grinding - Spiralbohrerschleiflehre
garnet, Granat
garnet cloth, Granatschleifleinen
garnet combination, Granatkombination
garnet paper, Granatpapier
gauge, Lehrwerkzeug, Lehre, Anschlag
gauge, to messen, lehren
gauge, caliper - Grenzrachenlehre
gauge, dial - Meßuhr
gauge, go - Minuslehre
gauge, grinding - Schleiflehre für hinterdrehte Fräser
gauge, internal caliper - Lochlehre
gauge, limit - Grenzlehre
gauge, micrometer - Mikrometer
gauge, not go - Pluslehre
gauge, plug - Lehrdorn
gauge, reference - Kontrollehre
gauge, ring - Lehrring
gauge, snap - Rachenlehre
gauge, stop - Anschlag
gauge, twist drill grinding - Spiralbohrerschleiflehre
gear, Zahnrad, Getriebe
gear, back - Vorgelege
gear, bevel - Kegelrad
gear, change - Wechselrad
gear down, to ins Langsame übersetzen
gear, helical - Pfeilrad, Schraubenrad
gear, herringbone - Pfeilrad
gear, internal - Zahnrad mit Innenverzahnung
gear ring, Tellerrad
gear, sliding - Schieberad
gear, spur - Stirnrad
gear up, to ins Schnelle übersetzen
gem grinding, Edelsteinschliff
general purpose stone, Abziehstein für allgem. Zwecke
general shope use, allgem. Werkstatts-Zwecke
generating method, Abwälzverfahren
get out of true, to unrund werden
gib, Führungsleiste
gib, taper - Keilleiste
glass, Glas
glass paper, Glaspapier
glaze, to glanzschleifen, polieren, glätten
glazed wheel, „blank" gewordene Scheibe
glazer, Polierscheibe, Polierplatte
glazing, Verschmieren, Glattwerden
gloss, Politur, Glanz
gloss, to polieren
glossiness, Politur, Glätte
glossy, sehr glatt
glue, Leim
glue, to leimen, zusammenleimen
goggles, Schutzbrille
gouge stone, Hohlmeißelstein
grade, Härte, Grad, Güte, Sorte, Qualität
grade scale, Härteskala
grade, to abstufen, ordnen
grading, Härtebestimmung, Sorteneinteilung
grading report, Härtebericht
graduation, Gradteilung
grain, Korn, Körnung
grain grading-department, Kornsieberei
grain marks, Kornmarken
grained, gekörnt
grained fine - feinkörnig
grained -, coarse grobkörnig
grained, medium - mittelfeinkörnig
granular, körnig
granulate, to körnen
granulation, Körnung
grease, Fett
grease, to schmieren

green, (grün) neu
green grit wheel, Grünkorn-Schleifscheibe
green wheel, neue Scheibe, ungebrannt
grind, to schleifen, polierschleifen
grind coarse, to grobschleifen
grind cooly, to Mühl schleifen
grind dry, to trockenschleifen
grind even, to ebenschleifen
grind first, to vorschleifen
grind internally, to innenschleifen, ausschleifen
grind preparatory, to vorschleifen
grind rough, to vorschleifen, grobschleifen
grind out, to ausschleifen
grind valves, to einschleifen (Ventile)
grind wet, to naßschleifen
grinder, **Schleifer,** Schleifmaschine
grinder, automatic sizing - automat. Kaliberschleifmaschine
grinder belt, Schleifband, Schleifriemen
grinder, belt - Bandschleifmaschine
grinder, center - Spitzenschleifmaschine
grinder, crank shaft - Kurbelwellenschleifmaschine
grinder, cutter - Fräserschleifmaschine, Werkzeugschleifmaschine
grinder, cylindrical - Rundschleifmaschine
grinder, disc - Flächenschleifmaschine
grinder, drill - Bohrerschleifmaschine
grinder, emery wheel - Schmirgelmaschine
grinder, foot - Schleifmaschine mit Fußantrieb
grinder hand - Handschleifmaschine
grinder, hoe - Hackenschleifmaschine
grinder, internal Innenschleifmaschine, Buchsenschleifmaschine
grinder, knife - Messerschleifmaschine
grinder, lap - Schleifmaschine (zum Blankscheuern von Lötflächen)
grinder, open side - Einständer-Flächenschleifmaschine
grinder, plain - einfache Schleifmaschine
grinder, plane surface - Flächenschleifmaschine
grinder, planer type surface - Planschleifmaschine
grinder, plow - Pflugscharschleifmaschine
grinder, portable - transportable Schleifmaschine
grinder, power - Schleifmaschine mit Kraftantrieb
grinder-rest, Schleifmaschinen-Lünette
grinder, sickle blade - Sichelblatt-Schleifmaschine
grinder, scissors - Scherenschleifer
grinder stand, Schleifplatz
grinder, stove plate - Ofenplattenschleifmaschine
grinder, surface - Flächenschleifmaschine
grinder, toolpost, Ständerschleifmaschine
grinder, tool - Werkzeugschleifmaschine
grinder, universal - Universalschleifmaschine
grinder, wet - Naßschleifmaschine
grinder, wheel - Räderschleifmaschine
grindery, Schleiferei
grinding, Schleifen, Polieren
grinding allowance, Schleifzugabe
grinding angle, Schleifwinkel
grinding and buffing attachment, Schleif- und Poliervorrichtung
grinding apparatus, Schleifapparat
grinding arbour, Schleifdorn
grinding art, Schleifkunst
grinding attachment, Schleifvorrichtung
grinding attachment, bevel - Konischschleifvorrichtung
grinding attachment, center - Körnerspitzen-Schleifapparat
grinding attachment, crankshaft - Kurbelschleifeinrichtung
grinding attachment, cylindrical - automatisches Schleifen
grinding belt, Schleifband
grinding bench, Schleifbank, Schleifstuhl
grinding carriage, Bettschlitten mit Schleifsupport
grinding, coarse rohschleifen
grinding compound, Schleifpaste, Einschleifpaste
grinding, cone - Kegelschleifen
grinding cylinder, Schleifring
grinding, cylindrical - Rundschliff, Zylinderischschleifen
grinding, decorative verzierschleifen
grinding department, Schleiferei
grinding department, tool - Werkzeugschleiferei
grinding, dry - Trockenschliff
grinding, dull - mattieren, mattschleifen
grinding edge, Schleifkante

grinding equipment, Schleifzubehör, Schleifausrüstung
grinding, even - ebenschleifen
grinding expert, Schleiffachmann, Schleifsachverständiger
grinding face, Stirnseite der Scheibe
grinding, fine verfeinern
grinding, flat surface - Planschliff
grinding, gauge - Lehrenschliff
grinding, gem - Edelsteinschliff
grinding, general - allgemeine Schleifzwecke
grinding glass, Schleifglas
grinding grain, Schleifkorn
grinding, hand - Handschleifen, Handschliff
grinding head, Schleifkopf
grinding, in - einschleifen
grinding, internal - Ausschleifen, Hohlschleifen, Innenschleifen
grinding lathe, Schleifbank
grinding material, Schleifmaterial
grinding machine, Schleifmaschine
grinding machine, automatic automatische Schleifmaschine
grinding, machine - Maschinenschliff
grinding machine, ball - Kugelschleifmaschine
grinding machine, belt Bandschleifmaschine
grinding machine, bench - Bankschleifmaschine
grinding machine, caliper gauge - Lehrenschleifmaschine
grinding machine, cam - Nockenschleifmaschine
grinding machine, center - Spitzenschleifmaschine
grinding machine, centerless, Spitzenlose Rundschleifmaschine
grinding machine, circular - Rundschleifmaschine
grinding machine, cut-off Abstech-Schleifmaschine
grinding machine, drum - Trommelschleifmaschine
grinding machine, drill and pointing Spiralbohrerschleifmaschine
grinding machine, emery - Schmirgelschleifmaschine
grinding machine, floor - Fußbodenschleifmaschine
grinding machine, headstock - Schleifmaschinen, Spindelstock
grinding machine, internal - Innenschleifmaschine, Hohlschleifmaschine
grinding machine, pearl - Perlenschleifmaschine
grinding machine, piston ring - Kolbenringschleifmaschine
grinding machine for plane irons, Hobelmesser-Schleifmaschine
grindnig machine pneumatic - Preßluftschleifmaschine
grinding machine, portable - tragbare Schleifmaschine
grinding machine, roll - Walzenschleifmaschine
grinding machine, sandpaper - Sandpapierschleifmaschine
grinding machine, saw - Sägenschärfmaschine
grinding machine, semi-automatic - halbautomatische Schleifmaschine
grinding machine, shear - Scherenschleifmaschine
grinding machine, stone - Steine-Schleifmaschine
grinding machine, surface - Flächenschleifmaschine
grinding machine, track - Schienenschleifmaschine
grinding machine, universal tool - Universal-Werkzeugschleifmaschine
grinding machine head stock, Schleifmaschinenspindelstock
grinding mark, Schleifmarke
grinding method, Schleifverfahren
grinding mill, Schleiferei, Schleifmühle, Schleifanstalt, Schleifkotten Schleife
grinding operation, Schleifvorgang,
grinding out, Ausschleifen
grinding out grooves, Rillenschleifen
grinding out plane -, Planschleifen
grinding plate, Schleifplatte
grinding position, Schleiflage, Schleifstellung
grinding powder, Schleifpulver
grinding practice, Schleifpraxis
grinding precision, Maßschleifen
grinding production, Mengenschleifen
grinding pressure, Schleifdruck
grinding rest, Schleifsupport, Schleifauflage, Schleifmaschinensupport
grinding roll, Schleifwalze

grinding, roll Walzenschleifen
grinding room, Schleifraum
grinding shop, Schleifwerkstatt
grinding, smooth, egalisieren
grinding surface, Schleiffläche
grinding surface Planschleifen
grinding, spherial Kugelschleifen
grinding, spot Zentrischleifen
grinding time, Schleifzeit
grinding to gauge, maßschleifen
grinding tool, Schleifzeug
grinding, wet - Naßschliff
grinding wheel, Schleifscheibe, Schleifrad
grinding wheel arbour, Schleifscheibendorn
grinding wheel dresser, Schleifscheiben-Abdrehvorrichtung
grinding wheel factory, Schleifscheibenfabrik
grinding slide Schleifscheibenschlitten
grinding wheel spindle, Schleifscheibenspindel, Schleifspindel
grinding with set-up wheel, Pließten
grinding work, Schleifarbeit
grindstone, Schleifstein
grindstone box, Schleifsteintrog
grindstone dust, Schleifstaub
grindstone swarf, Schleifschmand
grindstone trough, Schleifsteintrog
grit, Körnung
grit size, Korngröße
groove, Nut, Rinne, Eindrehung, Einschnitt
grooving, Nuten, mit Furchen versehen
ground, clearance - Hohlgeschliffen
ground, dead - mattgeschliffen (Glas)
ground, joint - geschliffener Stoß
ground, machine - mit der Maschine geschliffen
grow tired, to stumpf werden
guard, Schutzhaube, Schutzbügel
guide, Führung
guide, bar grinder - Schleifmaschine mit Führungsstange
guide ways, Führungsbahnen
gummer, saw - Maschine oder Scheibe zum Schleifen von Sägezähnen
gumming, Verharzen

band grinder, Handschleifmaschine
hand grinding, Handschliff
hand lapping, Handläppen
handed, right and left - rechts- und linksgängig
handle, Handgriff
hardened, gehärtet
hardness, Härte
handness, degree of - Härtegrad
hardness test, Härteprüfung
hardness testing machine, Härteprüfmaschine
head, Kopf, Kopfstück
headstock, Spindelstock
headstock swivel base, Spindelstock-Drehplatte
headstock center, Spindelstock-Spitze
headstock index finger, Spindelkopf-Winkelstellungsanzeiger.
havy duty disc, schwere Leinenscheibe
heavy duty grinder, Schleifmaschine für schwere Stücke
heavy duty bench grinder, Bankschleifmaschine für schwere Stücke
heel scouring paper, Absatzschleifpapier
high polish, to hochglanzpolieren, hellputzen
high production grinding, Massenschliff
high speed wheel, schnellaufende Scheibe
high tensile metal, Metall hoher Festigkeit
hob cutter, Schneckenfräser,. Abwälzfräser
hoe grinder, Hackenschleifmaschine
hold true its shape, to die Form bei behalten (Schleifscheibe)
holder, diamond tool - Diamanthalter
hole, Loch, Bohrung
hollow, hohl, Hohlkehle, Bohrung
hollow, to grind - hohlschleifen
hollow ground, hohlgeschliffen
hone, Abziehstein
hone, to abziehen
honing, Honen, Reibschleifen, Ziehschleifen
honing equipment, Honeinrichtung
honing head, Honahle, Ziehschleiffahle
honing stick, Honstein
hood, Schutzkappe, Schutzhaube
hook hone - stone, Anglerstein (Hakenstein)
HP (horse power) required, Kraftbedarf
housing, Gehäuse, Kasten, Ständer
hub, Nabe
hydraulic, hydraulisch

impart a cutting edge, to einem Werkzeug eine scharfe Schneide verleihen
imprint polishing, prägepolieren
increase, to zunehmen
in-feed, Beistellung, Zustellung
in-feed method, Einstech-Verfahren
in-feed grinding method, Tiefenschliff-Verfahren
index, Zeiger, Anzeiger
index, to schalten, teilen
indicator, dial - Meßuhr
inserted teeth saw, Sägescheibe mit eingesetzten Zähnen
inspect, to prüfen
inspection, Kontrolle, Abnahmeprüfung
instrument hone, Abziehstein für ärztliche Instrumente
integral, aus einem Stück
interchangeable, austauschbar
interlock with, to verriegeln
interlocking splice, Fingerform-Verbindung (Schleifband)
intermittent, aussetzend
internal grinder, Innenschleifmaschine, Büchsenschleifmaschine
internal grinding, Ausschleifen, Innenschliff
internal grinding attachment, Innenschleifeinrichtung
internal grinding fixture, Innenschleifvorrichtung
internal grinding machine, Innenschleifmaschine, Hohlschleifmaschine
ironcenter, Eisen-Kern
ironing, Feinschleifprozeß (Steinindustrie)
ironing wheel, Sonderscheibe (Glättscheibe) zum Schleifen flacher Steinoberflächen
joint, Verbindung
joint, puckered - faltige, runzlige Verbindung
joint, ground - geschliffener Stoß
jointer stone, Tischler-Stein
journal, Drehzapfen

keen edge, scharfe Schneide
keenness, Schärfe
keenness, angle of - Schneidwinkel
keep keen, to scharfhalten
kerosine, kerosene, Petroleum
keyseat, Keilnut
kiln, Brennofen
kitchen stone, Küchenstein
knee, Winkeltisch, Konsole
knife grinder, Messerschleifmaschine, Schleifmaschine für Hobelmesser
knife grinding, Messerschleifen
knife sharpener, Messerschärfer
knife stone, Maschinenmesserstein
knock off, to ausrücken
knock out, to auswerfen

label, Aufschrift, Etikett, Marke, Zettel
lacquer, Lack
lacquer, to lackieren
lacquer polishing machine. Lackpoliermaschine
lacquer sanding, Abschleifen von Lackanstrichen mit Schleifpapier
land, Rippe, Phase (an Werkzeugen)
lap, to läppen, schleifen, schmirgeln, polieren
lap, flaches Schmirgeleisen, Schmirgelgerät
lap, box board - Holzpolierscheibe
lap, diamond - Diamantschleifwerkzeug
lap, plug - Schleifdorn
lapping, Läppen, Sauberschleifen, Staubschleifen, d. h. mit pulverförmigen Mitteln unter Verwendung von Kluppen oder Scheiben aus Gußeisen, Stahl usw.
lapping machine, Läppmaschine
lathe center grinding apparatus, Drehbankspitzen-Schleifapparat
lathe chuck, Drehfutter
lathe, grinding - Schleifbank
lathe, grinding and buffing - Schleif- und Polierbank
lathe, screw polishing - Schraubenkopfpolierbank
lathe room, Dreherei (Fertigdreherei der gebrannten Scheiben)
lathers' stone, Axtstein
layer, Schicht, Belag
lazy wheel, schlecht schneidende Schleifscheibe
length to ground, Schleiflänge
lengthwise, der Länge nach
level, Ebene, waagerecht
leger, Hebel
light duty machine, leichte Arbeitsmaschine Maschine leichter Bauart
lime, Vienna - Wiener Kalk
limit, Grenze, Toleranz
limit gauge, Toleranzlehre, Grenzlehre

limit, high and low - oberes und unteres Grenzmaß
limit of accuracy, Genauigkeitsgrenze
line, to be in - with, ausgerichtet zu
line, to - up with, ausfluchten
link, Kulisse, Gelenk
lip, cutting - Schneidkante
lip of a tool, Schneide
live spindle driving pulley, Antriebsscheibe für das Werkstück
live spindle locking pin, Mitnehmerspindel-Feststellstift
load, Belastung
loaded wheel, zugesetzte, verschmierte Scheibe
loading of a wheel, verschmieren einer Scheibe
height of centers, Spitzenhöhe
load, up, to sich zusetzen
locate, to einstellen, zentrieren
locating device, Aufspann-, Haltevorrichtung
locati, Einstellung, Zentrierung
lock, to, festspannen, verriegeln
longitudinal reverse of the grinding wheel, Längsvorschub der Schleifscheibe
loose grain, loses Korn
loosen, to abspannen, losspannen
looseness, Spiel
loss, idle - Leergangsverlust
lost motion, toter Gang
low tensile metal, Metall niedriger Festigkeit
lubricant, Kühlmittel
lubricate, to, schmieren
lubrication, Schmierung
lubricator, Schmiereinrichtung
lump form, Stückform, Klumpen
luster, lustre, Glanz

machine, burnishing - Poliermaschine
machine, burring - Abgratmaschine
machine, centerless grinding - Spitzenlose Rundschleifmaschine
machine, cutting off - Abstechmaschine
machine-foundation, Maschinenfundament, Maschinenaufstellung
machine-grinding - Maschinenschliff
machine, grinding - Schleifmaschine
machine, lapping - Läppmaschine
machine-lapping, Maschinenläppen
machine, polishing - Poliermaschine
machine, riddling - Siebmaschine
magnesite, Magnesit, Talkspat
magnetic chuck, magnetisches Spannfutter
magnetic holding device, Magnetische Spannvorrichtung
make of machine, Fabrikat
mandrel, Dorn
manufactured abrasive, künstliches Schleifmittel
mark, center - Körnerpunkt
mark, chatter - Rattermarke, Zittermarke
mark, off, to anreisen, vorzeichnen
masterblock, Vergleichstück
match, to in Eingriff sein mit, kämmen mit
material, grinding - Schleifmaterial
measure, Maß
measure, to, messen
measurement, to take a - eine Messung machen
mechanism, Getriebe
mesh, Siebweite, Maschenweite
mesh sieve, Maschensieb
metcalf wheel dresser, Abdrehvorrichtung (Scheibe)
micrometer, Mikrometer
microscope, shop - Werksatt- Mikroskop
mill, grinding - Schleiferei
mixer, Mischmaschine
mixing machine, Mischmaschine
mixing room, Mischraum
modul, Modul eines Zahnrades
moisture proof, feuchtigkeitsbeständig
motion, Bewegung
motion, feed - Vorschubbewegung, Schaltung
motion, link - Kulissensteurung
motion, lost - toter Gang
motion, quick - Schnellgang
motion, rapid traverse - Schnellgang
motion, reciprocated - hin- und hergehende Bewegung
motion, reverse - Rücklauf
motor, adjustable speed - Stufenmotor
motor, constant speed - Motor mit konstanter Geschwindigkeit
motor, reversing - Umsteuermotor
motor, variable speed - Motor mit veränderlicher Geschwindigkeit
mould, to formschleifen (Marmorindustrie)
moulded shapes, geformte Streifen aus Schleifpapier
moulding, Formschleifen

moulding knife grinding machine, Kehlmesser-Schleifmaschine
mounted points, montierte Spitzen (auf Spindeln)
mounting of wheel, Montieren der Scheibe
movement, Bewegung, Getriebe
multiblade, mehrschneidig
multiple drum sanding, Schleifen auf Mehrzylinder-Schleifmaschine

natural abrasive, natürliches Schleifmittel, natürlicher Schleifrohstoff
natural stone, Naturstein
nick, Kerbe, Scharte, Riefe
non-positive, nachgiebig
non-pulling edge, nicht kratzende Schneide
nose, spindle - Spindelkopf
notch, to einkerben
nozzle, water - Wasserdüse (für Naßschliff)
nut, adjusting - Stellmutter

oblique, schräg
oblong, länglich
offhand grinding, Schleifen von Hand
offset, Ansatz, Vorsprung (an einer Schleifscheibe)
off-set, versetzt, gekröpft
oil film, Oelfilm
oil, lubricating -, Schmieröl
oil, machine - Maschinenöl
oil, non fluid - konsistentes Fett
oilstone, Oelstein
o. k., in Ordnung, abnahmefähig
opaque by grinding, to make mattschleifen
open coating, offene Streuung
open eoated abrasive paper, Schleifpapier mit offener Streuung
open side surface grinder, Einständer-Flächenschleifmaschine
operate, to bedienen
operating speed, Arbeitsgeschwindigkeit
operation, Arbeitsgang
operator, Arbeiter, Bedienungsmann
oscillate, to pendeln
oscillating curve grinding machine, Oszillierende Kurvenschleifmaschine
outfit, Ausrüstung
outline, Umriß, Profil
outline, surface - Mantellinie

output, Leistung einer Maschine
oven, Brennofen
overhang, Ueberhang, Ausladung
overhead grinding machine, hängende Schleifmaschine
overhung grinding wheel, fliegende Schleifscheibe
overtime, Ueberstunden
overtravel, Ueberweg

pace, Arbeitstakt, Rhythmus
pad, Bimskissen, Schleifkissen
pad, flexible schmiegsame Schleifleinenscheibe
pad, pressure, Druckauflage
pan, oil - Oelfangschale
paper, Papier
paper, abrasive -, Schleifpapier
paper, emery - Schmirgelpapier
parallel, Parallelstück (zum Aufspannen von Werkstücken
Paris red, Pariser Rot, Polierrot
paß to - inspection, abnehmen (bei der Revision)
pasta, Paste
pattern, Muster, Modell
pawl, Sperrklinke
pawl, ratched -, Schaltklinke
pedestal buffer, Ständerschwabbelmaschine
pencil points for engraver's chuck, Scheibspitzen für Graveurstichel
periphery, Umfang
peripheral speed, Umfangsgeschwindigkeit
phenol-resinoid, Kunstharz (Scheibenbindemittel)
phenol-resinoid bonded wheel, Schleifscheibe in Kunstharzbindung
pickle, to beizen
piece, work - Arbeitsstück
pillar, Ständer (einer Maschine)
pilot, Führungszapfen
pin, Stift, Bolzen, Zapfen
pin wheel, Kleine Schleifpapier- oder Schleifleinenscheibe (Schuhindustrie)
pin, taper - Keilstift
pin, to befestigen, verdübeln
pinion, Ritzel
pipe, Rohr
piston ring grinding machine, Kolbenringschleifmaschine

piston grinder, Kolbenschleifmaschine
pitch, Steigung, Teilung
pivot, Drehzapfen
pivoted, drehbar
plain, einfach, glatt, zylinderisch
plain back rest, einfacher Setzstock
plain cam grinding machine, einfache Nockenschleifmaschine
plain grinder, einfache Schleifmaschine
plain straight wheel, einfache glatte Scheibe
plain wheel, gerade Scheibe
plane iron Hobelmesser
plane grinding, Planschleifen
plane iron cap, Hobeleisen-Klappe
planer type surface grinder, Planschleifmaschine
planish, to glätten, polieren, schlichten
plant, grinding - Schleifanstalt
plate, Platte, Scheibe, Blech, Tisch
plate, grinding - Schleifplatte
plate, to plattieren
play, Spiel, Luft
play, end - toter Gang
pliessten, schleifen, polieren
plow grinder, Pflugschar-Schleifmaschine
plug, Dorn, Zapfen
plug, lap - Schleifdorn
plungecut grinding, Formschliff, Tiefenschliff
pneumatic grinding machine, Preßluft-Schleifmaschine
pocket hone, Taschenmesser-Abziehstein
point, konische Feile
polish, Politur
polish, brilliant or high - Hochglanzpolitur
polish, to polieren, polierschleifen, grobpolieren, glanzschleifen, glätten, glänzen
polish, clear klarschleifen
polish dry, to trockenpolieren
polish finish, to fertigpolieren
polisher, Polierer, Polierplatte, Polierstab
polish grinding, Schleifpolieren
polish, pre - Vorporlieren
polish, smooth - schlichtschleifen
polish, tumbling - kugelpolieren
polishing, Politur, Grobpolieren, Polieren
polishing agent, Poliermittel
polishing barrel, Poliertrommel
polishing brush, Schleifbürste
polishing clamp, Polierbüchse (mit Halter)
polishing cloth, Polierleinen
polishing cone, Polierdorn
polishing department, Polierabteilung
polishing drum, Poliertrommel
polishing felt, Polierfilz
polishing file, Polierfeile
polishing, first vorpolieren
polishing hammer, Polierhammer
polishing, imprint prägepolieren
polishing iron, Polierstahl, Poliereisen, Glättzahn
polishing machine, Poliermaschine
polishing material, Poliermateral
polishing pad, Polierkissen
polishing powder, Polierpulver
polishing rouge, Polierrot
polishing shop, Poliererei
polishing stone, Polierstein
polishing stick, Polierholz, Polierfeile
polishing, second - nachpolieren
polishing surface, Polierfläche
polishing table, Poliertisch
polishing tool, Polierwerkzeug
polishing wax, Polierwachs
polishing work, Polierarbeit
porosity, Porosität
porous, porös, durchlässig
portable grinder, transportable Schleifmaschine, Handschleifmaschine
positive, starr, unnachgiebig, zwangläufig
pot wheel, Topfscheibe
pounce, Bimsstein-Pulver
pounce, to mit Bimsstein abreiben
pouncing paper, Feinschleifpapier, Pouncingpapier
pour a wheel, to, eine Scheibe ausgießen
powder, Pulver
powder, emery - Schmirgelpulver
powder, polishing - Polierpulver
power grinder, Schleifmaschine mit Kraftantrieb
precision grinding, Präzisionsschliff, Maßschleifen
precision wheel, Präzisions-Scheibe
pregrinding paste, Vorschleifmasse
preliminary grinding, Vorschliff
pre-polish, vorpolieren
pressed wheel, gepresste Scheibe
pressfinish, Aufkugeln, Presspolieren
press mould method, Pressverfahren
press polishing, Polierdrücken
process, Verfahren
profiled wheel, profilierte Scheibe
profiling attachment, Profilier-Vorrichtung
protecting cap, Schutzhaube

protecting cover, Schutzhaube
protecting guard, Schutzhaube
protecting hood, Schutzhaube
protection hood, Schutzhaube
protection flanges, Schutzflanschen
protector for grinding wheel, Schleifscheiben-Schutzkappe
puddled process, Gießverfahren
puddled wheel, gegossene Scheibe
pull, belt - Riemenzug
pulley, Riemenscheibe
pulling, **Kratzen** (einer stumpfen Schneide)
pulp grinding wheel, Holzschliffscheibe (Papierfabr.)
pulp stone, Schleifscheibe aus Sandstein od. künstlich hergestellt für Holzschliff (Papierfabrikation)
pumicate, to bimsen (polieren mit Bimssteinpulver auf der Pappelholzscheibe in der Glasindustrie)
pumice, Bimsstein
pumice brick, Bimsstein
pumice stone, Bimsstein
pump, geared - Zahnradpumpe
pump, control oilgear - Flüssigkeitsgetriebe
purifying treatment, Reinigungsbehandlung
purity, Reinheit
putty of emery, Schmirgelbindemittel
putty powder, Zinnasche, Zinnoxyd
pyrometric cone, Segerkegel

quality of finish, Güte des Fertigschliffes
quarry, Steinbruch
quartz, Quarz
quartzsand, Quarzsand
quartzy, quarzig
quick adjusting lever, Schnelleinstellhebel

rack, Zahnstange
rack, tool - Werkzeuggestell
rag wheel, Schwabbelscheibe
rail joint grinder, Schienenstoßschleifmaschine
raised dovetail, erhöhter Schwalbenschwanz (Schleifscheibenform)
rake, Brust- oder Spanwinkel
range, feed and speed - Vorschub- und Geschwindigkeitsbereich
ratchet, Zahngesperre
rate, Geschwindigkeit, Maßstab, Verhältnis
rate, piece - Stücklohn
ratio, gear - Uebersetzungsverhältnis
rattler, Putztrommel
raw material, Rohmaterial
ray of sparks, Funkengarbe
ray wheel, Schwabbelscheibe
razor hone, Rasiermesserstein
razor strop, Streichriemen
razor strop dressing, Streichriemen-Paste
reamer, Reibahle
reaper file, Feile für Mähmaschinenmesser
rebush, to wieder ausbleien
recess, Aussparung, Vertiefung
recess wheel, Schleifscheibe mit Aussparung
recessed, ausgespart
reciprocate, to hin und her bewegen
recoat a wheel, to eine Scheibe neu belegen
reducing socket, Reduziereinsatz
reface, to einschleifen (Ventile)
refillable stone holder, auswechselbarer Steinhalter
re-finish, nachpolieren
register, to einstellen, feststellen
regrind, to neuschleifen, nachschleifen
regulate, to regulieren
regulating wheel, Transportscheibe, Regulierscheibe, Regelscheibe
relief grind, to hinterschleifen
relief grinding attachment, Hinterschleifeinrichtung
remove stock, to Material entfernen, Span abnehmen
repeat order, Nachbestellung
re-polish, wiederpolieren
reservoir, oil - Oelbehälter
rest, back - Lünette, Setzstock
rest, grinder - Schleifmaschinelünette
rest, grinding - Schleifauflage, Schleifsupport
rest, hand - Handauflage. Handsupport
rest, slide - Werkzeugschlitten, Kreuzsupport
rest, steady - Setzstock
rest, swivel- drehbarer Schlitten

return, quick - Schnellrückgang
reversible, umsteuerbar
reversing lever, Umsteuerhebel
revolution, Umdrehung
revolutions per minute (r. p. m.) minutliche Umlaufzahl
revolve, to sich drehen
rigid machine, starre Maschine
rigidity, Starrheit, Stabilität
ring, clear - reiner Klang (bei der Klangprobe)
rod, Stange, Stab
rod, end measuring - sphärisches Endmaß
roll, Walze, Rolle
roll grinding, Walzenschleifen
roll grinding machine, Walzenschleifmaschine
roll scouring block, Schleifblock zum Schleifen von Walzen
roller, feed - Zuführungswalze
rolling, Rollieren
room, tool - Werkzeuglager, Werkzeugausgebe, Werkzeugmacherei
rottenstone, Englische Erde, Tripel
rouge, Polierrot
rough, to schruppen, rohschleifen, schruppschleifen
rough, to grind - vorschleifen, grobschleifen
rough - lapping, Vorläppen
roughing, Schruppen, Grobschliff
roughing cut, Schruppen
roughing operation, Vorschliff
round edge gouge stone, Hohlmeißelstein mit abgerundeter Kante
round faced wheel, Scheibe mit abgerundetem Rand
round stick, Rundfeile
roundness, Rundheit
roundness, out of - unrund
r. p. m. (revolutions per min.), minutliche Umlaufzahl
rub, to reiben, polierschleifen
rub down, to abschleifen
rub with emery, to schmirgeln
rub with pumice stone, to bimsen
rubber, Abziehstein, Rutscherstein, Handschleifstein
rubber, Gummi, Guttapercha
rubber bond, Gummi-Bindung
rubber bonded wheel, Scheibe in Gummibindung
rubber grinding wheel, Gummi-Schleifscheibe
rubber wheel, Gummischeibe
rubbing brick, Rutscher
rubbing block, Rutscher
rubbing down, Abschleifen
rubbing stone, Rutscherstein
rubs, Rutscher
rule, Lineal, Maßstab, Zollstock
run, to laufen
run to - true, genau rund laufen
running, steady - ruhiger Lauf
running test, Laufprobe
rupture of the grinding wheel, Bruch oder Bersten der Schleifscheibe
rust, Rost
rustless, rostfrei

safe guard, Schutzvorrichtung
safety hood, Schutzhaube
safety speed test, Probelauf der Scheibe
safety device, Schutzvorrichtung
safety flange, Sicherheitsflansch
sager, Brennkapsel
sagers' clay, Kapselton
sand, Sand
sand blast, Sandstrahlgebläse
sand blast apparatus, Sandstrahlgebläse
sand paper, Sandpapier
sand paper grinding machine, Sandpapier-Schleifmaschine
sand papering machine, Sandpapiermaschine
sand stone, Sandstein
sander cone, konische Leinenscheibe
sander, disc - Sandpapierschleifmaschine
sanding belt, Schleifband
sanding liquid, Schleifwasser
sanding paper, Schleifpapier
sanding machine, Sandpapiermaschine
sanding wet, Naßschliff
satin finish, mattgeschliffen
saucer wheel, Tassenscheibe
saw cutting machine, Sägenschärfmaschine
saw dust, Sägemehl
saw grinding machine, Sägenschleifmaschine
saw gummer, Sägenschleifscheibe
saw gumming, Sägenschärfen
saw gumming wheel, Sägenschärfscheibe

saw sharpening machine, Sägenschärfmaschine
scale of hardness, Härteskala
scissors grinder, Scherenschleifmaschine
scobs, Sägespäne
score, to kerben
scouring, Scheuern, Blankputzen
scotch trowel, Polierblech
scraper-brunisher, Polier-Schaber
scraping method, Schabeverfahren
scraping tool, Schaber
scratch, to, kratzen
scratch, Kratzer
screen, Sieb
screen, to sieben
screen numbering, Sieb-Numerierung
screw polishing machine, Schraubenkopf-Poliermaschine
screw tap, Gewindebohrer
scythe stone, Sensenwetzstein
seam, Grat, Naht
seam, welded Schweissnaht
sectional wheel, Segmentscheibe
segment, Segment
section, cross - Querschnitt
secure, to befestigen
segment, Segment
segmental grinding wheel, Segment-Schleifscheibe
segmental wheel chuck, Segmentscheiben-Spannfutter
selection of bond, Auswahl der Bindung
self contained grinder, eingekapselte Schleifmaschine, Schleifmaschine mit in sich abgeschlossenem Antrieb
self feed, Selbstgang
self generating method, Abwälzverfahren
semi-automatic, halbselbsttätig
set, Satz, Reihe
set, to setzen, aufspannen, einstellen, einziehen, schränken
set up, to belegen, einrichten
set - up time, Herstellungszeit
set-up wheel, Pliesst-Scheibe
setting up, Aufstellung, Einrichten, Montage
sever, to, trennen
s. f. p. m. S. F. P. M. (surface feed per minute), Umfangsgeschwindigkeit in Fuß pro Minute
shaft, Spindel, Welle
shaft, cam — Nockenwelle
shaft, crank — Kurbelwelle
shaft, drinving — Antriebswelle
shaft, line — Hauptwelle, Antriebswelle
shaft, splined — Nutwelle
shape a wheel, to, einer Scheibe eine Form geben
shape of face, Randform
shape of wheel, Scheibenform
shaping device, wheel - Schleifscheibenabdrehvorrichtung
shaping shop, Vordreherei für gegossene Rohscheiben
sharp, scharf
sharpen, to schleifen, schärfen
sharpening the teeth square across, Gradschliff der Zähne
sharpening the teeth to alternating angles, Schrägschliff der Zähne
sharpening device, Schärfvorrichtung
sharpening machine, saw - Sägenschärfmaschine
sharpening stone, Abziehstein, Schärfer
sharpness, Schärfe
shaving machine, Rohlingschlichtmaschine (Scheibenherstellung)
shear grinding machine, Scherenschleifmaschine
sheepskin, Schafleder
sheer cut, sauberer, reiner Schnitt
sheet, Blatt, Blech, Bogen
sheet goods, Blattware
shell, Hülse, Napf, Buchse, Zarge
shellac, Schellack
shellac bond, Schellackbindung
shellac bonded wheel, Scheibe in Schellackbindung
shock, free from stoßfrei
shoemakers' sharpening stone, Schuhmacherabziehstein
shoemakers' strop, Schuhmacherstreichriemen
shop, Werkstatt
shop, jobbing - Reparaturwerkstatt
sieve, Sieb
sieve, to sieben
sift, Sieb
sift, to sieben
sifter, Siebmaschine
sifting machine, Siebmaschine
silica, Kieselsäure
silicate process wheel, Scheibe in Silikatbindung

silicon carbide, Siliziumkarbid
silversmith' stone, Abziehstein für Silberarbeter
single drum sanding, Schleifen auf Einzylinder-Schleifmaschine
single pulley drive, Einscheibenantrieb
sinter, to Sintern
size, Mass
size, actual, Ist-Mass
size basic, Nennmaß
size ob hole, Lochgröße
sizing device, Meßeinrichtung
sizing grinder, Kaliberschleifmaschine
sleeve, Muffe, Buchse, Hülse
sleeve, taper - Reduziereinsatz
slide, Schlitten, Schieber, Gleitbahn
slide, cross - Querschlitten
slide, grinding wheel, Schleifscheibenschlitten
slide, to gleiten
slide, tool Werkzeugschlitten
sliding caliper, Schieblehre
sliding table, Tisch-Schlitten
slip, Handschleifstein
slip ring, Schleifring
slip stone, Abziehstein in Messerklingenform
slip, to gleiten, rutschen
slot, Schlitz, Nut
slot cutter, Schlitzfräser
slotting, Vertiefen, Vertiefung
slow down the wheel speed, to Die Scheibengeschwindigkeit herabsetzen
slow up the wheel speed, to die Scheibengeschwindigkeit heraufsetzen
slow motion, Langsamgang
smooth, glatt
smooth, dead spiegelglatt, poliert
smooth, to glätten, schlichten
smooth edge, scharfe Schneide
smooth grinding, egalisieren
smooth ground surface, glatt geschliffene **Oberfläche**
smooth polish, schlichtschleifen
smoothing, Feinschleifen, Feinmachen, Glattschleifen (3. Arbeitsgang beim Glasschleifen)
smoothing tool, Schlichtstahl
snagging, Guß putzen
snagging machine, Gußputzmaschine
snagging wheel, Gußputzscheibe, Abgratscheibe
snake marks, Schlangenmarken
socket, Kegelhülse, Einsatz, Futter
sodium chloride, Kochsalz
soft bond, Weiche Bindung
solid, masiv, aus einem Stück
solution, Lösung
solution, grinding - Schleiflösung
sound the wheel, to die Scheibe durch Klang prüfen
spacing, Teilungsabstand
spanish white, Schlämmkreide
spar, Spat (Mineral)
spark, to feuern, funken
spark out, to (die Schleifscheibe) ausfeuern lassen
sparks, ray of Funkengarbe
speed, Geschwindigkeit
speed box, Getriebekasten
speed, cutting - Schnittgeschwindigkeit
speed indicator, Umdrehungszähler
speed of cut, Schnittgeschwindigkeit
speed, peripheral - Umfangsgeschwindigkeit
special shaped wheel, Scheibe in Sonderform
spherical grinding, Kugelschleifen
spherical wheel, kugelige Scheibe
spindle, Welle, Spindel
spindle box, Schleifspindellager
spindle, grinding wheel - Schleifwelle
spindle speed, Spindelgeschwindigkeit
spiral, Schnecke, Spiralkurve, spiralig
spiral gear grinder, Schraubenrad-Schleifmaschine
splice, Verbindungsstelle
splice-dovetail, schwalbenschwanzförmige Verbindungsstelle
spline shaft grinding machine, Kurbelwellenschleifmaschine
spot grinding, Einstechschleifen
spring, Feder
spur gear grinding machine, Stirnwand-Schleifmaschine
square, Winkel, Viereck, Quadrat
square stick, Vierkantfeile
square up, to rechtwinklig bearbeiten
stainless steel, rostfreier Stahl
stand, wheel - Schleifbock
standard, Norm, Normal
standard, Ständer, Säule
standard gauge, Normallehre
standard shape, Normalform
stationary, feststehend
steel, burnisher Polierstahl

steel center - wheel, Stahlkernscheibe
steel shot, Stahlsand
steel, tool - Werkzeugstahl
steel wool, Stahlwolle
steel, stainless - nichtrostender Stahl
step, Stufe
stepped, four - vierstufig
stepped, geometrically - geometrisch abgestuft
stiched cotton buff, genähte Schwabbelscheibe aus Baumwolle
stick, Feile
stick, burnishing - Polierstahl
stick, dresser - Abdrehapparat mit Abdrehstab
stick, emery - Schmirgelholz
stick, round - Rundfeile
stich, square Vierkantfeile
stick, triangular Dreikantfeile
stick, to sich festsetzen, anheften
stock feed, Materialvorschub
stock, foot - Reitstock
stock, head - Spindelstock
stock - removal, Material-Abnahme
stone, Stein
stone, emery - Schmirgelstein
stone grinding machine, Steine-Schleifmaschine
stone, oil - Ölstein
stone, pumice - Bimsstein
stone, to, polieren, nachschleifen
stoneholder, refillable, auswechselbarer Steinhalter
stoning, Reibschleifen
stop, Anschlag
stoppering work, Glasstöpsel-Einschleifen
storage of wheel, Lagerung der Scheibe
stove plate grinder, Ofenplatten-Schleifmaschine
straight, gerade, zylinderisch
straight-in and straight-trough, Einstech- und Durchlaufverfahren (auf der spitzenlosen Rundschleifmaschine)
straight-in feed method, Einstechverfahren
straight cup wheel, gerade Topfscheibe
straight wheel, gerade Scheibe
straightness, Geradheit
strap polisher, Polierleder
stress, Spannung, Beanspruchung
stroke, Hub
stroke, return - Rücklauf
structure, Struktur, Gefüge
stud, Bolzen, Stift, Kopfschraube
supply, oil - Ölzufuhr, Ölzuleitung
support, work - Werkstückauflage
surface, Fläche, Oberfläche
surface, to Oberfläche abschleifen
surface, to glattschleifen, (Betonböden)
surface grinder, Flächenschleifmaschine
surface grinding machine, Planschleifmaschine, Flächenschleifmaschine
surface tenacity, Oberflächen-Festigkeit
surfacing, Glätten (von Betonböden)
surfacing machine, **Flächenschleifmaschine**
swarf, grinding - Schleifschmant
swarf, wheel - Schleifsteinpulver
swing frame grinder, Schwingrahmen-Schleifmaschine
switch, Schalter
swivable, schwenkbar, drehbar
swivel, Schwenkplatte
swivel table, Tisch-Schwenkplatte
swivel table knob, Tisch-Schwenkschraube
swivel, to schwingen, schwenken
synthetic resin, synthetisches Harz
synthetic resin bond, synthetische Harzbindung

table, Tisch (einer Maschine)
table reversing dog, Tischumkehranschlag, Umsteueranschlag
table area, Tischfläche
table handfeed, Tischvorschub von Hand
table scale, Tischskala
table surface grinder, Tisch-Flächenschleifmaschine
table travel control, Tischlaufsteuerung
table traverse, Tischbewegung
tachometer, Umdrehungszähler
tag, Etikett
take out nicks, to Scharten entfernen
tallow, Talg
tandem grinder, Doppelschleifmaschine
tap the wheel, to die Scheibe anschlagen
tap, Gewindebohrer
taper, Konus, Konizität, Verjüngung
taper hole, konische Bohrung
tapered, konisch
tapered wheel, konische Scheibe
temper, Härtegrad, Vergütung
temper, to härten, anlassen, tempern

template = **templet**, Schablone
tension, Spannung
tension adjusting knob, Spitzendruck-Adjustierschraube
tensile strength, Zerreißfestigkeit, Zugfestigkeit
test, Prüfung, Probe
test, spark - Funkenprobe
test, to prüfen
testing apparatus, Prüfapparat
testing machine, Prüfmaschine
texture, Struktur, Gefüge
thickness, Stärke, Dicke
thickness rim, Wandstärke, Randstärke (bei Topfscheiben)
tickness back, Bodenstärke (bei Topfscheiben)
thread, Gewinde
thread gauge, Gewindelehre
thread grinding, Gewindeschliff
thru-feed, zylindrisch Schleifen im spitzenlosen Schleifverfahren
thrust, end - Gegendruck
tighten, to anziehen, spannen
tightener, belt - Riemenspanner
time, idle or lost, Leerlauf
tolerance, Toleranz
tool, Werkzeug
tool, burnishing - Polierwerkzeug
tool dresser, Werkzeugschleifer
tool grinder, Werkzeugschleifmaschine
tool grinding, Werkzeugschliff
tool grinding department, Werkzeugschleiferei
tool grinding machine, Werkzeugschleifmaschine
tool holder, Werkzeughalter
tool machine, Werkzeugmaschine
tool mark, Werkzeugmarke
tool post grinder, Ständerschleifmaschine
tool room, **Werkzeuglager**, Werkzeugausgabe, Werkzeugmacherei
tool room grinding, Schleifen in der Werkzeugmacherei
tool, scraping - Schaber
tool shop, Werkzeugmacherei
tool, spoon - Polierlöffel
tooth, Zahn
tooth rest, Zahnauflage
toothing, Verzahnung
touch up, to leicht anschleifen
tough bond, zähe Bindung
toughness, Zähigkeit

train of gears, Räderwerk, Übersetzung
transmission, Übertragung
transverse motion, Quergang (des Arbeitstisches)
transverse slide, Querschlitten
travel of the grinding wheel, Längsvorschub der Schleifscheibe
traverse of the wheel, Vorschub der Scheibe
traverse speed, Vorschubgeschwindigkeit
treadle, Fußhebel
treating, heat - Warmbehandlung
treatment, heat - Warmbehandlung
trial run, Probelauf
triangular stick, Dreikantfeile
trim, to abrichten (Schleifscheibe)
trim off, to abgraten
trimming the emery wheel, Abziehen der Schmirgelscheibe
trimming the grindling wheel, Ablehren des Schleifsteines
trip, to auslösen (durch Knagge)
tripping, feed - Vorschubauslösung
tripoli powder, Tripel
trough, Behälter, Trog
trowel, scotsch - Polierlöffel (Formerwerkzeug)
true grinding wheel, schlagfreie Scheibe
true, out of - unrund
trueness, Genauigkeit
true, to abrichten
truing — up, Abrichten
truing device, Schleifscheibenabrunder
truing the grinding wheel, Ablehren der Schleifscheibe
try, cut and ausprobieren
tub wheel, Topfscheibe
tube, Rohr
tube, feed - Zuführungsrohr (für Material
tubing, Rohrleitung
tumbler, Putztrommel
tumbling, Trommeln, Kugelpolieren
tumbling barrel, Putztrommel
tumbling polish, kugelpolieren
tungsten, Wolfram
tungsten carbide, Wolframkarbid, Carboloy, Widia-Metall
tungsten carbide tool, Wolframkarbid-Werkzeug
turn, to drehen
turn into shape, to die Scheibenform vordrehen
turret, Revolverkopf

twin, doppel . . ., Zwillings . . .
twin wheel grinding, Das Schleifen mit Doppelschleifscheiben
twist drill center grinding attachment Spiralbohrerspitzen-Schleifapparat
twist drill grinding gauge, Spiralbohrer-Schleiflehre
twist drill relief grinding machine, Spiralbohrer-Hinterschleifmaschine
type of grinding, Schleifart

unbushed, unausgebleit
unclamp, to abspannen (ein Werkstück)
uncoated, unbelegt
uncoated paper, unbelegtes Papier
uneven wear, ungleichmäßige Abnutzung
uneveness, Ungleichheit, Ungleichmäßigkeit
unfinished, unbearbeitet, unappretiert
unfinished stick, unappretierte Feile
unfired mixture, ungebrannte Mischung
uniform, gleichmäßig
uniformity, Gleichmäßigkeit
uniformity of grit, Gleichmäßigkeit der Körnung
universal chuck, Universalspannfutter
universal grinder, Universalschleifmaschine
universal tool and cutter grinder, Universal-Werkzeugschleifmaschine
universal tool grinder, Universal-Werkzeugschleifmaschine
universal tool grinding machine, Universal-Werkzeugschleifmaschine
untrue, unrund
untrue, to become unrund werden
up-keep, Instandhaltung
upright, Ständer
urea formaldehyd resin, Harnstoffharz

valve, Ventil
valve grinding, Ventilschleifen
valve grinding compound, Ventilschleifpaste
valve refacer, Ventilschleifmaschine
varnish, to, lackieren
velocity, Geschwindigkeit
vent, Entlüftung
vernier, Nonius
vibration, Erschütterung
vice, Schraubstock
vitrified bond, keramische Bindung
vitrified process wheel, keramisch gebundene Scheibe
vitrified wheel, keramisch gebundene Scheibe
vitrify to. versintern, brennen
vitrifying kiln, Brennofen

wabbly wheel, schlagende Scheibe
walrus leather, Walroßleder
warping, Werfen der Scheibe
washer, Etikett, Unterlegscheibe, Zwischenlage
washer, felt - Filzunterlegscheibe
washer, rubber - Gummischeibe
washers of blotting paper, Pappscheiben
waste, Abfall
water attachment, Wasserzuführungseinrichtung
water, gneard support, Halter für den Wasserfänger
water hone, Wasserstein
water level, Wasserstand
water piping, Wasserzuführgsrohr
water proof paper, wasserdichtes Schleifpapier
water supply, Wasserzuführung
ways, Führungen
wear, Abnutzung, Verschleiß
wear, to — (out) sich (vollständig) abnutzen
wear out of true, to sich ungleichmäßig abnutzen, unrund werden
webbed, verrippt
wedge shaped, keilförmig
wet grinder, Naßschleifmaschine
wet grinding, Naßschliff
wet grinding attachment, Naßschleifeinrichtung
wet rubbing operation, Naßschliff mit wasserfestem Schleifpapier
wet tool grinder, Werkzeug-Naßschleifmaschine
wheel, Scheibe, Schleifscheibe
wheel assortment, Scheibenzusammenstellung
wheel balancing equipment, Scheibenauswucht-Vorrichtung
wheel breakage, Scheibenbruch
wheel, buffing - Polierscheibe
wheel, carriage, Schleifschlitten

wheel dresser, Abdrehwerkzeug, Abziehwerkzeug
wheel dressing device, Scheibenabdreh-Vorrichtung
wheel dressing stick, Scheiben-Abdrehstein
wheel driving pulley, Schleifscheiben-Antriebsscheibe
wheel edge, Scheibenrand
wheel face, Stirnseite der Scheibe, Randform
wheel, formed - Formscheibe
wheel grinder, Räderschleifmaschine
wheel, grinding Schleifscheibe
wheel guard, Schleifscheiben-Schutzhaube
wheel head, Scheibenkopf
wheel life, Scheibenlebensdauer
wheel press, Scheibenpressung
wheel, ray - Schwabbelscheibe
wheel, regulating - Transportscheibe
wheel replacement, Scheibenersatz
wheel set-up, aufgeleimte Polierscheibe
wheel shape, Scheibenform
wheel slide, Schleifbock
wheel speed, Scheibengeschwindigkeit
wheel stand, Schleifbock
wheel standplate, Schleifbock-Drehplatte
wheel stand slide, Schleifbock-Schieber
wheel stand swivel, Schleifbock-Drehplatte
wheel, straight - gerade Scheibe
wheel structure, Scheibengefüge
wheel, to charge a etwas auf eine Scheibe auftragen
wheel truing attachment, Scheibenabricht-Vorrichtung
wheel wear, Scheibenabnutzung
wheel, wet Naßschleifscheibe
whet, to schleifen, wetzen
whet - stone, Abziehstein, Wasserstein
whetting, Wetzen
whitening, Schlämmkreide
width, Breite
width of face, Bandstärke
wire, Draht
wire-web wheel, Drahtgewebe-Scheibe (mit Drahtgewebe-Einlage)
wood bushing, Holzbuchse
work, Werkstück
work between centers, Spitzenarbeit
work, chuck - Spannfutterarbeit
work, duplicate - Reihenfertigung
work holder, Werkstück-Halter
work locating device, Werkstück-Zentriervorrichtung
work, repetition - Reihenfertigung
work speed. Werkstückgeschwindigkeit
worm, Schnecke
worm grinding machine, Schneckenschleifmaschine
worm, multiple thread mehrgängige Schnecke
wrench Schraubenschlüssel

Anhang

Zusammenstellung

von

Sonder-Schleifmaschinen

Supplement

Assortment

of

Special Grinding Machines

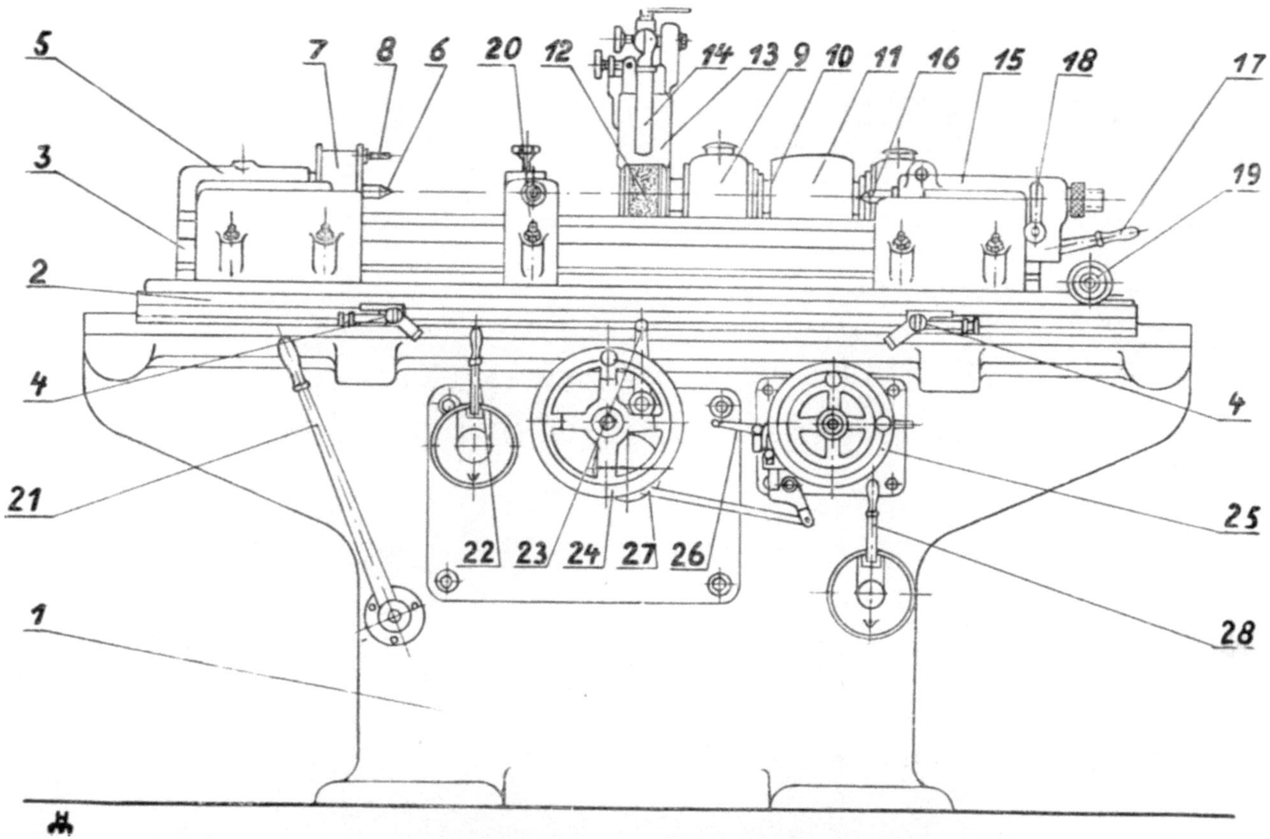

Die einfache Rundschleifmaschine
The Plain Cylindrical Grinding Machine

Die einfache Rundschleifmaschine

The Plain Cylindrical Grinding Machine

1. **Maschinengestell** bed, body, frame
2. **Untertisch (Tischschlitten)** sliding table
3. **Obertisch (Tischschwenkplatte)**
4. **Tischumkehr-Anschläge** table reversing dogs
5. **Spindelstock** headstock
6. **Spindelstock-Spitze** live center
7. **Spindel-Antriebsscheibe** live spindle driving pulley
8. **Mitnehmer** driver
9. **Schleifspindellager** wheel spindle box
10. **Schleifscheibenspindel** grinding wheel spindle
11. **Schleifscheiben-Antriebsscheibe** wheel driving pulley
12. **Schleifscheibe** grinding wheel
13. **Schutzvorichtung** protection hood; wheel guard
14. **Wasserzuführung** water piping
15. **Reitstock** tailstock; footstock
16. **Reitstock-Spitze** dead center
17. **Schnelleinstellhebel für die Spitze** quick adjusting lever
18. **Spitzen-Feststellhebel** clamping lever
19. **Schrägverstellung des Tisches** swivel table knob
20. **Brille oder Lünette** center rest
21. **Ein- und Ausrückhebel** starting and stopping lever
22. **Einstellhebel für Tisch-Vorschübe** traverse speed lever
23. **Umsteuerhebel** table reversing lever
24. **Handrad für Tischvorschub** hand wheel for table feed; cross-feed hand wheel
25. **Handrad zum Beistellen der Schleifscheibe** hand wheel for grinding wheel feed
26. **Feinzustellung der Schleifscheibe** fine in-feed of the grinding wheel
27. **Selbsttätige Zustellung der Schleifscheibe** automatic in-feed
28. **Steuerhebel für Werkstück-Geschwindigkeiten** work speed control lever

The Plain Cylindrical Grinding Machine

Die einfache Rundschleifmaschine

1. **bed body frame** Maschinengestell
2. **sliding table** Untertisch (Tischschlitten)
3. **Obertisch (Tischschwenkplatte)**
4. **table reversing dogs** Tischumkehr-Anschläge
5. **headstock** Spindelstock
6. **live center** Spindelstock-Spitze
7. **live spindle driving pulley** Spindel-Antriebsscheibe
8. **driver** Mitnehmer
9. **wheel spindle box** Schleifspindellager
10. **grinding wheel spindle** Schleifscheibenspindel
11. **wheel driving pulley** Schleifscheiben-Antriebswelle
12. **grinding wheel** Schleifscheibe
13. **protection hood; wheel guard** Schutzvorrichtung
14. **water piping** Wasserzuführung
15. **tailstock; footstock** Reitstock
16. **dead center** Reitstock-Spitze
17. **quick adjusting lever** Schnelleinstellhebel für die Spitze
18. **clamping lever** Spitzen-Feststellhebel
19. **swivel table knob** Schrägverstellung des Tisches
20. **center rest** Brille oder Lünette
21. **starting and stopping lever** Ein- und Ausrückhebel
22. **traverse speed lever** Einstelhebel für Tisch-Vorschübe
23. **table reversing lever** Umsteuerhebel
24. **hand wheel for table feed; cross-feed hand wheel** Handrad für Tischvorschub
25. **hand wheel for grinding wheel feed** Handrad zum Beistellen der Schleifscheibe
26. **fine in-feed of the grinding wheel** Feinzustellung der Schleifscheibe
27. **automatik in-feed** Selbsttätige Zustellung der Schleifscheibe
28. **work speed control lever** Steuerhebel für Werkstück-Geschwindigkeiten

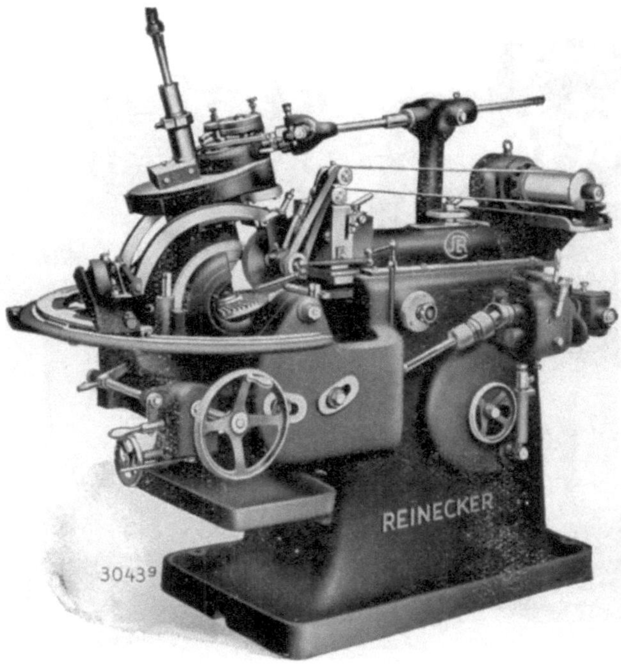

Automatische Kegelradschleifmaschine zum Schleifen von Kegelrädern mit geraden oder spiralgewundenen Zähnen sowie von Hypoidkegelrädern (Abwälzverfahren)

Automatic bevel gear grinding machine for grinding bevel gears with straight or spiral (helical teeth and hypoid bevel gears (Self-generating method)

Automatische Innenschleifmaschine mit hydraulischem Antrieb zum Ausschleifen zylindrischer und konischer Bohrungen

Automatic internal grinding machine with hydraulic drive for inside grinding of cylinder and taper holes

Automatische Stirnradschleifmaschine mit großem, feststehenden Schleifrad (Abwälzverfahren)

Automatic spur gear grinder with large fixed grinding wheel (self-generating method)

Automatische Stirnradschleifmaschine (Abwälzverfahren)

Automatic spur gear grinding machine (Self-generating method)

Automatische Stirn- und Schraubenrad-Schleifmaschine mit schwenkbarem Schleifkopf (Abwälzverfahren)

Automatic spur and spiral gear grinder with swivelling grinding head (Self-generating method)

Lehrenschleifmaschine

Caliper gauge grinder

Kolbenringschleifmaschine zum Schleifen der flachen Seiten an Kolbenringen-Federn und dergl.

Circular face grinding machine specially meant for grinding the faces of piston rings, springs etc.

Rund- und Einstechschleifmaschine mit hydraulischer Tischbewegung und hydraulicher Schnellverstellung des Schleifbocks

Cylindrical and plunge-cut grinding machine with hydraulic table feed and hydraulic rapid movement of thee wheel carriage

Rundschleifmaschine mit selbsttätiger Längsbewegung des Schleifschlittens auf besonderem Bett

Cylindrical grinding machine with automatic traverse of wheel carriage on side bed

Horizontale Planschleifmaschine

Horizontal face grinding machine

Innen- und Planschleifmaschine mit hydraulischem Tischantrieb

Internal and face grinding machine with hydraulic table drive

Einfache Nockenschleifmaschine

Plain cam grinding machine

Flächenschleifmaschine, einständrig, mit vertikaler Schleifspindel

Surface grinding machine with one fixed housing and with vertical grinding spindle

Flächenschleifmaschine mit horizontaler Schleifspindel

Surface grinding machine with horizontal grinding spindle

Flächenschleifmaschine mit horizontaler Schleifspindel

Surface grinding machine with horizontal grinding spindle

Flächenschleifmaschine, doppelständrig, mit vertikaler Schleifspindel

Surface grinding machine with two fixed housings and with vertical grinding spindle

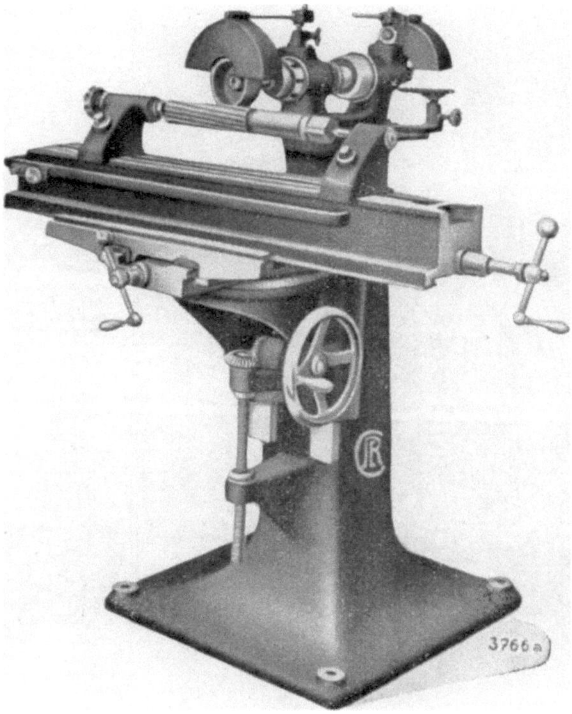

Werkzeugschleifmaschine zum Schärfen von Fräsern aller Art, Reibahlen und Gewindebohrern usw.

Tool grinding machine specially meant for sharpening milling cutters of all shapes, reamers, taps, etc

Spiralbohrer-Hinterschleifmaschine

Twist drill relief grinding machine

Universal-Rundschleifmaschine mit Antrieb durch Deckenvorgelege oder Einzelmotoren.

Universal cylindrical grinding machine with drive by cuntershaft by different single motors

Universal-Werkzeugschleifmaschine zum Schärfen von Fräsern aller Art, Reibahlen, Gewindebohrern usw.

Universal tool grinder specially meant for grinding milling cutters of all shapes, reamers, taps etc.

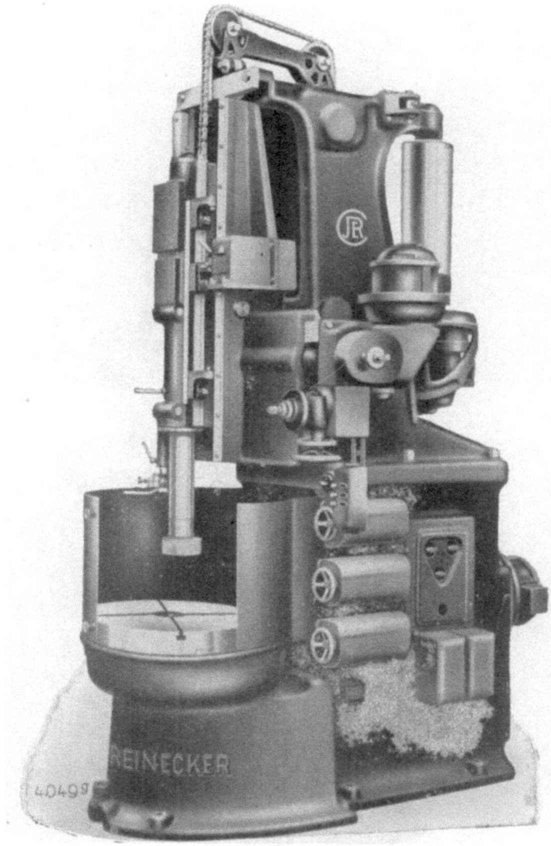

Vertikale Innenschleifmaschine zum Ausschleifen zylindrischer Bohrungen in Zylindern, Büchsen und dergl. mit rotierender Planscheibe für das Werkstück

Vertikal internal grinding machine for grinding the inside of hollow cylinders, bushings etc., with rotary circular table

Vertikale Innenschleifmaschine mit kreisender Spindel

Vertical internal grinding machine with planetary movement

Vertikale Innenschleifmaschine mit kreisender Spindel und Einrichtung zum Schleifen gekrümmter Kulissen.

Vertical internal grinding machine with planetary movement and oscillating attachment for grinding curved links

Schneckenschleifmaschine

Worm grinding machine

Bei Fragen zur Produktsicherheit wenden Sie sich bitte an:
If you have any questions regarding product safety,
please contact:

Walter de Gruyter GmbH
Genthiner Straße 13
10785 Berlin
productsafety@degruyterbrill.com